GAIA

The Practical Science
of Planetary Medicine

GAIA

The Practical Science of Planetary Medicine

James
Lovelock

OXFORD
UNIVERSITY PRESS

OXFORD
UNIVERSITY PRESS

Oxford   New York
Athens   Auckland   Bangkok   Bogotá   Buenos Aires   Calcutta
Cape Town   Chennai   Dar es Salaam   Delhi   Florence   Hong Kong   Istanbul
Karachi   Kuala Lumpur   Madrid   Melbourne   Mexico City   Mumbai
Nairobi   Paris   São Paulo   Singapore   Taipei   Tokyo   Toronto   Warsaw

and associated companies in
Berlin   Ibadan

First published in the United Kingdom in 1991 by
Gaia Books Limited
66 Charlotte Street
London, W1P 1LR
and
20 High Street
Stroud, Glos GL5 1AZ

First published in the United States by Harmony Books, a division of Crown Publishers, Inc.,
201 East 50th Street, New York, NY 10022. Member of the Crown Publishing Group

First published in paperback in the United States in 2000
by Oxford University Press, Inc.
198 Madison Avenue, New York, New York 10016

Oxford is a registered trademark of Oxford University Press

Library of Congress Cataloging-in-Publication Data is available
Lovelock, J. E.
[Healing Gaia]
Gaia: the practical science of planetary medicine/James Lovelock.
p.   cm.
First American ed. published under title: Healing Gaia. New York: Harmony Books, ©1991.
Includes bibliographical references
ISBN 0-19-521674-1
1. Geobiology—Philosophy.   2. Gaia hypothesis.
3. Biosphere.   4. Environmental protection.
5. Nature conservation.   I. Title.
QH343.4.L69   2000
508'.01—dc21   00-037497

1   3   5   7   9   8   6   4   2
Printed in Singapore
on acid-free paper

**To
my beloved wife
Sandy**

Frontispiece: Chalk cliffs, like these in Dorset, are typical along the southern coast of Britain. Similar deposits of limestone, calcium carbonate, occur in many parts of the world. They represent Gaia's assiduous work in the burial of carbon, being made up of the fossils of billions of microscopic marine organisms. These had used calcium carbonate in the construction of their shells.

Contents

Preface to the new edition

This book explores the Earth through the eyes of an imaginary planetary physician. I like to think of this physician as one who made house calls; someone living before the days of antibiotics and molecular biology; someone skilled at diagnosis, who could bring comfort, and could sometimes heal by guiding the process by which nature took its course.

The concept of planetary medicine implies the existence of a planetary body that is in some way alive, and can experience both health and disease. When this book was first published in 1991 the concept of a living planet, and Gaia theory from which it comes, was hotly disputed by many biologists who saw life as the singular property of living organisms and the genes they express. Feelings were strong and both Gaia as a theory, and Lynn Margulis and I as its proponents, were publicly discredited by a small but vociferous group of biologists and science writers. They attacked from the high ground of established science and, as in *Science*, 19th April 1991, denounced Gaia theory as: "unscientific", "dangerous", "pure fantasy". Most of all, they balked at the Earth being described as alive.

Now it is different and the last ten years have seen the widespread acceptance of what I like to think of as Gaian science and although the name Gaia is rarely used, scientists now talk freely of "Earth System Science". They recognize that a view of the Earth from the separated disciplines of biology, geology, and geophysics is not enough.

But as you read this book, keep in mind that Gaia is as yet unproven, and use it and the concept of planetary medicine as a way of seeing the Earth's problems differently.

It takes a lot of misunderstanding to upset a sensible scientist, therefore I accept that in the past Lynn Margulis and I were provocative and that I now need to make clear what we mean by "life". In this book I often describe the planetary ecosystem, Gaia, as alive, because it behaves like a living organism to the extent that temperature and chemical composition are actively kept constant in the face of perturbations. When I do I am well aware that the term itself is metaphorical and that the Earth is not alive in the same way as you or me, or even a bacterium. At the same time I insist that Gaia theory itself is proper science and no mere metaphor. My use of the term "alive" is like that of an engineer who calls a mechanical system alive to distinguish its behaviour when switched on from that when switched off, or dead. Engines on whose proper function many lives depend have health monitors; devices that ensure signs of failure are detected early enough for a cure, not a tragedy.

Should I, or you, be worried by these strictures from other scientists, some of whom are distinguished professors at famous universities? No, I don't think so. They were said in the heat of an argument now resoloved. Gaia is after all just another way of looking at the mysteries of the Earth.

Why should otherwise sensible scientists have lost their normal imperturbability over Gaia? I think it was a consequence of the long and exhausting war between biologists and creationists. In the heat of this battle scientists uncharacteristically became dogmatic and arrogant. I am an old-fashioned scientist who believes, as Freeman Dyson put it in his book *Infinite in all Directions*, that the ethic of science is based on a fundamental open-mindedness, a willingness to subject every theory to analytical scrutiny and experimental test. The Royal Society of London in 1660 proudly took as its motto the phrase *"Nullius in verba"* meaning "No man's word shall be final". There is no place for infallibility in science. I was also brought up to believe that science was serious but not sacrosanct and that creative science required a sense of wonder and a sense of humour.

Gaia theory may be wholly or partially in error. To a real scientist this is not as important as how well the theory fits these criteria: Is it useful? Does it suggest interesting experiments? Does it explain the puzzling data we have gathered? What are its predictions? Does it have a mathematical basis? Gaia gives positive answers to all these questions and now receives consideration from scientists who ten years ago regarded it with contumely.

Creative science is the province of working scientists, and few of these are found among our remaining critics. Those who still condemn Gaia as unscientific are, for the most part, science writers and professional science critics. Creative scientists are inspired by ideas that are often difficult to explain in words, but that nevertheless suggest experiments. The formal explanation almost always comes after the inspiration and the experiments. This is something that science writers and critics rarely wish to know, for it would make the telling of the stories so much more difficult.

This is not a contentious or argumentative book. It is not about whether Gaia theory is right or wrong but about using Gaia as a way to look at the Earth differently. I invite you to join with me and explore our planet with an imaginary planetary physician as your guide, someone who regards geophysiology, the systems science of the Earth, as the proper basis of his empiricism.

JAMES E LOVELOCK

Introduction

> "I speak as the representative, the shop steward, of the bacteria and the less attractive forms of life who have few others to speak for them. My constituency is all life other than humans."

The notion of a planet visiting the doctor is odd. It assumes for a start that the planet – in this case the Earth – is capable of being ill, and so is in some sense alive. It also assumes that there is a suitable doctor to visit – one with the knowledge and experience of planetary maladies to give sound advice. A physician, in fact, trained in planetary medicine. It is with the need for such a practice of planetary medicine that this book is concerned.

You may find it easier to envisage this hypothetical planetary medical check-up if you think back to the time when you last had a fit of hypochondria and imagined that you were the victim of some fatal but romantic disease. This usually happens after reading a medical article and identifying your minor symptoms with those described.

Society in the prosperous parts of the world is undergoing collectively a similar experience. The difference is that the apparent hypochondria is about the world itself instead of about our individual selves. The equivalent of the medical articles is the ever-growing flood of "doom" scenarios. There is no shortage of planetary diseases to identify with, from nuclear winter chills and greenhouse fevers to acid rain indigestion and ozone spots. These problems are real. But, as in hypochondria, we do not know whether such symptoms are the harbingers of disaster or no more than harmless growing pains of the planet.

Intelligent hypochondriacs do not consult a biochemist or a molecular biologist about their worries; they go instead to their doctor, a general practitioner of medicine. A good doctor knows that hypochondria often masks a real ailment different from the one imagined by the patient. Could it be that our deep hypochondria about the state of the global environment also masks a real disease of our planet? How can we find out, and who should we ask for advice? Just as a microbiologist might find it difficult to prescribe a cure for the common cold, it could be that the real plan-

The life of Gaia

Mats of microbial life cover the shallow coastal lagoons in Baja California, Mexico. In the self regulating life system of Gaia, microbes such as these play the leading role. Yet so deep is our human introspection that we care little about life other than ourselves – even though it sustains us all.

etary malaise is beyond the understanding of specialist scientists in fields like climatology or geochemistry. Is it wise then to seek only the advice of such experts? It may seem that we have no other options, for the practice of planetary medicine does not yet exist. How then should we go about establishing such a practice? What would be its tools and methods? What would be the qualifications of its professional practitioners? And what its scientific basis?

If the history of human medicine is a guide, planetary medicine will grow from guesses and empiricism, from practical solutions to immediate problems, from common sense and good hygiene. And its scientific basis will be physiology, the systems science of living organisms – or rather, geophysiology, the systems science of the Earth.

Fortunately for human medicine, physiology was introduced early in its history and strongly influenced its progress. The recognition by the 16th-century physician, Paracelsus, that "the poison is the dose" is a physiological enlightenment understood immediately in medicine. (It has yet to be discovered by those environmentalists who still seek the pointless and unattainable goal of zero for pollutants.) The wisdom of medicine grew when William Harvey (1578–1657) discovered the circulation of the blood. James Hutton (1726–97), the father of the science of geology, was inspired by the similar circulation of the planet's water to speak of the Earth in physiological terms, which was then as now the ground and framework of medicine.

James Hutton was a polymath who included medicine among his scientific qualifications. It was natural that he should, like some wise old barn owl floating above a meadow, have taken a holistic or "top-down", physiological view of the Earth. Most scientists of the 20th century disregarded James Hutton's physiological approach to Earth science. The younger, expert sciences – whether microbiology or biogeochemistry – took a narrower, reductionist, or "bottom-up" view, studying details and processes within their fields. They recognized that life on Earth affects its environment as well as adapting to it, but they lost Hutton's great vision that saw life and its material environment as a single system.

This is why I still think that established science is ill-suited to cope with the problems of global change. I am not suggesting that we should cast aside our vast store of scientific knowledge, still less that we abandon scientific method. But I do suggest we should look again at the evidence science has gathered, and see if the physiological view can better explain and predict what will happen next. I acknowledge the Earth system scientists, and the International Geosphere Biosphere Program (IGBP) who are doing this, but they represent only a small part of Earth science. In the 18th and 19th centuries, the science of physiology underpinned the empiricism of human medicine and enabled the physician to navigate a way to health through scientific uncertainties. Today, its analogue Earth system science, or as I have called it, geophysiology, can provide a basis for the empirical practice of planetary medicine and help us choose a path toward health for the whole system of Earth.

Understanding Gaia

Before I explain this practice of planetary medicine further, I must introduce Gaia – the patient with whom the physician will be concerned.

Gaia is the name of the Earth seen as a single physiological system, an entity that is alive at least to the extent that, like other living organisms, its chemistry and temperature are self-regulated at a state favourable for its inhabitants.

I describe Gaia as a control system for the Earth – a self-regulating system something like the familiar thermostat of a domestic iron or oven. I am an inventor. I find it easy to invent a self-regulating device by first imagining it as a mental picture. The model in the mind is then used to construct a prototype, and the prototype refined by trial and error until the self-regulating device is reduced to practice. The odd thing is that a working invention is extraordinarily difficult to explain in words. In many ways Gaia, like an invention, is difficult to describe. The nearest I can reach is to say that Gaia is an evolving system, a system made up from all living things and their surface environment, the oceans, atmosphere, and crustal rocks, the two parts tightly coupled and indivisible. It is an "emergent domain"– a system that has emerged from the reciprocal evolution of organisms and their environment over the eons of life on Earth. In this system, the self-regulation of climate and chemical composition are entirely automatic. Self-regulation emerges as the system evolves. No foresight, planning, or teleology (suggestion of design or purpose in nature) are involved.

I first stated the idea of Gaia in public in 1971 (see Chapter 1) and shortly after this began a collaboration with the eminent biologist, Lynn Margulis, that has continued ever since. At first we explained the Gaia hypothesis in words such as "Life, or the biosphere, regulates or maintains the climate and the atmospheric composition at an optimum for itself." This definition was imprecise, it is true; but neither Lynn Margulis nor I ever proposed that planetary self-regulation is purposeful. Yet we met persistent, almost dogmatic, criticism that our hypothesis was teleological. New theories always and rightly meet with opposition. It is part of the natural selection by science. In the arguments over Gaia quite often the metaphor not the science was attacked. Metaphor was seen as a pejorative, something inexact and therefore unscientific. In truth, real science is riddled with metaphor. Is physics less scientific because quarks have charm and colour? Science grows from imaginary models in the mind and is sharpened by measurements that check the fit of the models with reality.

Gaia became visible through the new knowledge about the Earth gained from space and from the extensive investigations of the

Earth's surface, oceans, and atmosphere during the past few decades. While this view lends itself to poetic metaphor, it is also a hard science theory of our planet that came from a top-down view from space. The top-down view of the Earth as a single system, one that I call Gaia, is essentially physiological. It is concerned with the working of the whole system, not with the separated parts of a planet divided arbitrarily into the biosphere, the atmosphere, the lithosphere, and the hydrosphere. These are not real divisions of the Earth, they are spheres of influence inhabited by academic scientists.

Gaia theory is now under test. The evidence gathered from the Earth itself will decide whether or not it should be taken as scientific fact. Even if in the end Gaia should turn out to be no more than a metaphor, it would still have been worth thinking of the Earth as a living system. Such a model is fruitful: it has already led to many discoveries about the Earth that could not have come from conventional wisdom.

In this book I ask that you concede there might be something in the Gaia theory. To acknowledge Gaia at least for the purpose of argument. I do not expect you to become converts to a new Earth religion. I do not ask you to suspend your common sense. All I do ask is that you consider Gaia theory as an alternative to the conventional wisdom that sees the Earth as a dead planet made of inanimate rocks, ocean, and atmosphere, and merely inhabited by life. Consider it as a real system, comprising all of life and all of its environment tightly coupled so as to form a self-regulating entity. Perhaps you already have it in mind when you use that vague, rather ill-defined word "biosphere" that seems to have a different meaning for each occasion of its use. The word biosphere was originally coined by the Austrian geographer Seuss to mean the inhabited realm of the Earth and properly should mean this and nothing more.

I recognize that to view the Earth as if it were alive is just a convenient, but different, way of organizing the facts of the Earth. I am, of course, prejudiced in favour of Gaia and have filled my life for the past 25 years with the thought that the Earth might in certain ways be alive, not as the ancients saw her, as a sentient goddess with purpose and foresight, but more like a tree – a tree that exists, never moving except to sway in the wind, yet endlessly conversing with the sunlight and the soil. Using sunlight and water and nutrients to grow and change. But all done so imperceptibly that, to me, the old oak tree on the green is the same as it was when I was a child.

The need for planetary medicine

Sometimes in anguish, concerned environmentalists ask if humans have become a leukemia of the Earth. Are we an organism that broke with the great community of Eden or of Gaia? Are we now breeding without check until our numbers and our toxins compromise the whole body of the Earth? It might seem so but leukemic

What is planetary medicine?

When thinking about planetary problems, I find it helpful to compare the examination made by a human physician with that by a hypothetical planetary physician. What are the instruments used? What samples are taken for biochemical analysis or biopsy? The table here lists some of these comparative tools and methods.

During the Earth's history there have been periods of disturbance to the system and of discomfort for its inhabiting organisms. The impacts of large planetesimals, or the major change of atmospheric state when oxygen became dominant, are examples.

Today, the presence of humans and their pollutants is creating another great stress. It is not so far-fetched to look on such states of the whole system as a disease. Even if no more than a metaphor, such a view is down to earth and practical.

Engineers design machinery whose failure would threaten life; they include what they call "machine health monitors". The engine and rotor mechanisms of helicopters incorporate such a health service, which ensures that the first symptoms of impending failure are used to warn us, in time to prevent disaster. That no biologist would recognize a helicopter as a form of life does not invalidate the engineer's approach.

To design a health monitoring service for a machine, the engineer has to have a thorough practical knowledge of how it works as a whole system. A doctor needs a similar practical knowledge of how the human system works, but needs also to know about the processes peculiar to living systems – birth and death, growth and healing. A planetary physician needs the skills and practical wisdom of the doctor, as well as those of the engineer, in seeking to understand the health of Gaia.

Even if you do not concede that the system of life and its environment on Earth, which I call Gaia, is alive – alive like a tree – you will surely admit that it is more alive than a helicopter or any machine.

Tools and methods in human and in planetary medicine

Measurement	Human	Planet
Temperature	Clinical thermometer	Satellite radiometer
Blood pressure	Sphygmomanometer	Barometer
Breathing	Stethoscope	Atmospheric carbon dioxide and oxygen monitors
Biochemical tests	Blood and urine samples	Air, sea and soil samples
Biopsy	Tissue samples	Ecosystem studies

cells do not debate their destructive role, nor consider a change of behaviour that might curb their own numbers. True, our presence on Earth has already changed the land surfaces and the atmosphere adversely. If we believe this to be a planetary disease then we need planetary medical advice now. Practical advice on how to restore a state of health. We cannot wait for big science planetary medical research to find the cure; there is no time.

I remember asking climatologists at a meeting in Brazil in 1985 which would happen first, elimination of the forests of Amazonia, assuming the present rate of clearance continued, or the development of a scientific model that could accurately predict the climate of the region when the trees were gone. They replied that the trees would have been cleared before they could give their answer. It is not that science has failed us – we need it in the long run – but for now it is pointed in the wrong direction and its responses are far too slow.

Consider, for example, the problem of the greenhouse gases. Modern science, even with the most powerful equipment, cannot predict the consequences, even a few years ahead, of the accumulation of carbon dioxide, methane, and other gases (see Chapter 8). Common sense and experience tell us that to keep adding these gases to the overburdened air will probably lead to disaster. Since this book was written we have continued to increase carbon dioxide above the normal interglacial level of 280, to 368 parts per million. We have increased another greenhouse gas, methane, above its natural level of 650, to 1760 parts per billion. Both of these human-induced increases are as great as those which occurred naturally between the last glaciation and the warm period that followed. The Earth's surface temperature has risen about 1°C since 1800, the start of the industrial revolution and over the same period the sea level has risen by 5 cm. Nothing we can conceivably do in the next few years will alter the unstoppable increase. We do not have the time to start all over again. We are about to experience the consequences of what we have already done to the Earth.

I think that we should follow the example of our forebears when confronted with recurrent disasters: use our common sense and try empiricism. It does not have to be right scientifically to get results. Consider the Romans: they knew that living on wetlands was unhealthy. They thought disease came from the bad odours in the air, so they drained the swamps and the disease, malaria, went away. Had they spent money instead on entomology and microbiology, in the course of time they might have discovered the malaria parasite and the fact that it was transmitted by mosquitoes. But many more would have died or led diminished lives before relief finally came by the same answer; namely draining the swamps. The example of the Romans is repeated throughout human history. The great extension of life during the nineteenth century was often not as the result of advances in biology or biochemistry, but usually by the application of common-sense medicine and civil engineering. Cholera and typhoid killed millions in Europe until clean drinking water became available. Again the preventive action by the civil engineers came long before the discovery of the causative agents and even longer before the discovery of antibiotics for the cure.

We need this pragmatic approach now if we are to solve our planetary ills in time. We need planetary medicine. Its approach may be empirical, even at times unscientific, but it is all that we have. I am not proposing some kind of alternative science, the equivalent in medicine to acupuncture or homeopathy. But the mainstream of science has wandered too far from its natural course. I believe in science and my aim is merely to deflate the tumescence of macho big science and calm it down. How else would it listen to my case for planetary medicine? If scientists are to recognize the value of empiricism in the troubled times to come, they must first acknowledge the extent of their ignorance about the Earth.

The nature of science

I am not saying that we do not need organized science; only that we need to recognize its frailty as a human institution, that it is slow, and its record in handling immediate environmental problems is far from good. It tends to do only those things that scientists find easy to do and want to do anyway. It concentrates almost obsessively on minor matters that happen to worry the public, such as carcinogens in the environment, or on recondite phenomena in the atmosphere that are intensely interesting to scientists. I find it deplorable that so much scientific effort has gone to research in the upper atmosphere when the large issues, such as forest clearance and the atmospheric greenhouse, have been, relatively speaking, ignored. This disproportion was changed in the last decade but vigilance is needed to keep the balance.

Most citizens of the developed world are deluded by the belief that pouring money on science is the way to get results. Sometimes the large-scale engineering development of an established discovery or invention is the right course to take. Penicillin, atomic energy, genetic engineering, and computers, all these have been made available to the public, expensively, but economically. Such development is good but it is not science, it is the engineering reduction to practice of the science that went before. Real science, the wondering about how the world works and the design of simple experiments to test the theories that thus come to mind, is like its companion creative activity, art, and best done quietly and inexpensively. When next you are asked to donate to some charity supporting research into a cure for cancer, consider how in Britain nearly ten times more money is spent on cancer research than on all other medical research, and yet the goal of a cure is as far away as ever.

The progress of good science is slow and unpredictable and all too often waits upon the appearance of a key thought in the mind of a genius. The mere employment of a hundred new and brightly polished doctors of philosophy from great universities to tackle the problem of global change is most unlikely to achieve anything other than provide them with secure and comfortable employment.

Make no mistake, to understand the physiology of the Earth, how Gaia works, requires a top-down view, a view of the Earth as a whole system; if you like, as something alive. The limitations of the reductionist bottom-up approach in science can be illustrated by the environmental problem of a hypothetical illness affecting a rare animal species. A disease severe enough to threaten the species with extinction and about which little is known other than that it is a non-infectious hepatitis. The traditional reductionist approach would be to kill some of the animals, dissect their livers, and subject them to biochemical and histological analysis. The sacrifice of a few animals this way would be justified on the grounds that it is the only way to save the species. Contrast this with the top-down approach. The planetary physician would look first at the ecosystem of which the animals are a part to see if the illness is the consequence of a larger disorder, not just specific to the animals themselves. Our planetary doctor might find that a change of climate which brought more rain had encouraged the growth of moulds and fungi on the food plants of the animals and that the moulds produced aflatoxin, a potent liver poison. The objective always would be to keep the animal intact and alive while seeking to discover the nature of its malady and of any dysfunction in the ecosystem to which it belonged.

I hope that for our species as part of the planetary ecosystem, part of Gaia, it is not too late to apply the right scientific approach. We are now in a position similar to that of a diabetic before the discovery of insulin. A vast government-funded programme to find the cure for diabetes might have stumbled on insulin, but sadly not fast enough to save the diabetic. Humans on Earth may be like this diabetic; irreversible damage may be done to the system of which we are a part before we discover how to live with it.

Living with Gaia

In the world today there is a feeling like that before a coming war, or of the ominous calm that precedes a tropical hurricane. With a hurricane we know what to do before it comes, what precautions to take, where to go to escape disaster. For the changes that threaten the world now, we have no detailed guide, we can only guess what they will be. Change may come gradually, but more often in a stressed system it arrives in a sequence of abrupt events, stepping from one level to another. We are in for surprises, events that could not have been predicted.

Environmentalists, churches, politicians, and science, all are concerned about the damage to the environment. But their concern is for the good of humankind. So deep is this introspection that even now, few apart from eccentrics really care about other living organisms. The oft-stated objection to the rape of the forests is that they might include within them some rare plant that bears the cure for cancer, or that the trees fix carbon dioxide, and that if they are cut down we may no longer enjoy our privilege of private transport. None of this is bad, only stupid. We are failing to recognize the true value of the forest as a self-regulating sub-system that keeps the climate of the region, and to some extent the Earth, comfortable for life. Without the trees there is no rain, and without the rain there are no trees. We do not have to become saints, only to achieve enlightened self-interest. If we can do this by letting the forest grow and sustain itself we shall have acknowledged our debt to the rest of life on Earth. It may cost no more than the payments we now make for clean water, refuse collection, street clearing, and sewage disposal. It will vastly benefit our environment and without the payment – our due tax to Gaia – we shall suffer just as surely as would the people of a city that refused to pay its taxes.

Modern medicine recognizes the mind and body as part of a single system where the state of each can affect the health of the other. It may be true also in planetary medicine that our collective attitude towards the Earth affects and is affected by the health of the planet. Christian teaching has it that "the body is the temple of the soul" and that this alone is a good enough reason for leading a healthy life. I find myself looking on the Earth itself as a place for worship, with all life as its congregation. For me this is reason enough for doing everything that is in my power to sustain a healthy planet.

Perhaps because I think this way, I take special comfort from an unusual sacred place which I often visit with my wife, Sandy. It is the small church of St Michael de Rupe, perched upon the central peak of a long-extinct Miocene volcano, about half a mile south of the village of Brentor in Devon, some ten miles from my home at Coombe Mill.

At the peak of Brent Tor, on a fine day when the wind comes in from the broad Atlantic bringing clear fresh air, the green dappled fields and woods of Devon stretch out to a far horizon 30 miles away. They form a landscape that looks good, perhaps because the farming is still pre-agribusiness. This pleasant prospect encompasses three-quarters of the view and stands in contrast to the small mountain mass to the east called Dartmoor. Here is treeless brown tundra, that from a distance seems to have slumped like the tailing of a mine. It was a forest once, but Iron Age people destroyed it. There is always at the peak of Brent Tor a sense of sacredness, as if it were a place where God and Gaia meet. The feeling is intense, like that felt in great cathedrals, caverns, and on other mountain tops. Of course, it could be rationalized scientifically as a physical sense, as yet poorly understood, by which the brain processes the sound impinging on our ears and converts the multitude of signals into a "vision in sound" of our surroundings. Bats and other animals that see by sound must possess this sense. Maybe we do also, but only become aware of it in places such as Brent Tor. It is this sense that creates what my physicist friend, Peter Fellgett, calls ambience. Whatever the reason, Brent Tor and places like it have a sense of peace. They seem to serve as reference points of health against which to contrast the illness of the present urban or rural scene.

I have asked in this book that you stretch your imagination quite a distance. That you take on for discussion the idea that the Earth is in some sense alive and that the diagnosis and treatment of its ills become an empirical practice, planetary medicine. You may not wish to do this because you are sure that our planet was created by God specifically for the benefit of humankind, and that we have the right to use it as we wish. You may prefer to see humans as the stewards, good farmers and fishermen, taking care of the Earth for the benefit of people, particularly our descendants. You may be one of those scientists who are quite certain that the Earth is just a lump of dead rock moistened with water and surrounded by an atmosphere composed of simple gaseous chemicals, and who regard life as something that merely inhabits and uses the Earth. If you hold fixedly to any of these views, then this book is not for you. This book is only for those who are curious and concerned about the Earth and who, like me, do not know all the answers. It is for those who wonder whether we really are God's chosen species – or whether we are not, instead, simply the most destructive event in the Earth's biological history. Above all, it is for those who are concerned that the single-minded quest to save humankind conflicts with the greater need to sustain the Earth as a fit and comfortable place to live. If we lose our habitat, the system of life and its environment on Earth, Gaia, will go on. But humankind will no longer be a part of it.

A guide to planetary medicine

In this new book of medicine, the Earth is the patient. Let us forget human concerns, human rights, and human suffering, and concentrate instead on our planet, which may be sick. We are a part of this Earth and we cannot therefore consider our affairs in isolation. We are so tied to the Earth that its chills or fevers are our chills and fevers also.

This book is written like one of those home medical encyclopaedias that families refer to for information or reassurance when someone is ill, and experts are unavailable or unable to help. I hope it will serve as a manual of this kind, helping you better to understand the Earth in health and disease, or when some new or frightening phenomenon afflicts the region of the Earth where you live. And I hope it will provide a guide that will enable you to decide what you can do personally, now, to prevent or at least alleviate the damage that lies ahead.

Guides for human medicine, written so as to be accessible and comprehensible to lay readers, often include chapters on the scien-

Brent Tor

"There is always at the peak of Brent Tor a sense of sacredness, as if it were a place where God and Gaia meet. Brent Tor and places like it serve as reference points for health against which to contrast the illness of the present rural or urban scene."

tific basis of medical practice with detailed descriptions of how the human body works. I have organized this book of planetary medicine so that the chapters that follow discuss Gaia under the headings of anatomy, physiology, epigenesis (origins and life history), biochemistry, metabolism, and climate (temperature) regulation. Chapter 8 deals with the people plague (disseminated primatemia), while the Conclusion is concerned with treatment, and the future of our relationship with Gaia.

Medical books often give case histories to illustrate the course and nature of disease. In its long, 3.8-billion-year history, the Earth has experienced disorder and disease on a planetary scale. Included within most of the chapters are case histories of such pathologies, events that were temporarily disabling but from which Gaia recovered, chosen to illustrate the subject of each chapter. If we think of the Earth as an organism, one that can be healthy or diseased, then maybe we can see Gaia outlined in this constellation of planetary pathologies, and understand more about planetary medicine. From the study of pathology we learn the nature of disease and how to cure or prevent it. Through an understanding of the abnormal we delineate, like a silhouette cut in a sheet of paper, the much more incomprehensible normal state of health.

If our presence is indeed a threat to the Earth then we are in the unusual position of being both the agents of disease and the observers of its effects. Pathology is a fine context in which to argue such deep questions as are we God's chosen ones? Or are we a plague on the face of the Earth? But there is no need for value judgements; even in ourselves some disease is a route to health, or can save us from consequences worse than the malady itself. We need only recall the many diseases of childhood, minor infections that immunize against the serious risk of an attack in later life. On the larger scale a plague, which eliminates the greater proportion of a species, can be both a disaster or a boon. Indeed, the pathology of error is fundamental to evolution by natural selection. Only through error and change can evolution proceed. Chance brings the happy error that allows an organism to inhabit a new niche, a new environment. Moreover the rules of Gaia are such that organisms that harm their environment do not long survive. We would do well to understand this rule, which may have fatal consequences for our species. Fortunately, we as persons also evolve our characters by a natural process of error and correction: we are able to learn. And nothing teaches better than a near miss.

The essence of living green, of being a citizen of Gaia, is not a fretful puritanism. If we can think of ourselves as a part of a giant living organism and perhaps even a cause of its indigestion, then we may be guided to live within Gaia in a way that is seemly and healthy. Even thinking this way is an antidote to the fatalism of accepting the Earth as dead, with life as just a passenger.

CHAPTER ONE

Recognizing Gaia

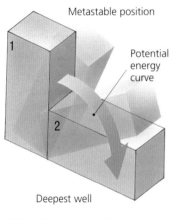

Metastable position

Potential energy curve

Deepest well

1 Disequilibrium (transitional equilibrium)

2 Equilibrium

What is equilibrium?

Equilibrium, or what scientists prefer to call "thermodynamic equilibrium", can best be described as a stable state from which no more energy can be extracted, as represented above by a brick lying on its long side (2): the brick has no further to fall. An unstable state of *disequilibrium* exists if energy is still available – as, for example, if the brick is stood on end (1). A small push is all that is needed to release the potential energy of motion of the brick, so that it falls to its equilibrium position. To a chemist, the state of equilibrium is closer to death than to contentment, since it is reached only when all free energy is used up. Nothing on this planet is at the equilibrium state.

For me, the personal revelation of Gaia came quite suddenly – like a flash of enlightenment. I was in a small room on the top floor of a building at the Jet Propulsion Laboratory in Pasadena, California. It was the autumn of 1965, the room was an office of the Biosciences Division, and I was talking with a colleague, Dian Hitchcock, about a paper we were preparing on a method for the remote detection of life on a planet.

In the early 1960s, NASA had sought my help in its quest to discover whether there was life on Mars. For the most part, the experiments proposed by my biologist colleagues seemed to me geocentric. They were seeking to find on Mars the life they were familiar with in their laboratories here on Earth. Better, I suppose, to do this than to seek a speculative form of life, but surely there was a truer and more certain way to detect Martian life? Earlier that year I had proposed, in a letter to *Nature*, some physical tests for the presence of planetary life. One of these involved taking a "top-down" view of the chemistry of the whole planet, rather than a local search at the site of landing. The test was simply to analyse the chemical composition of the planet's atmosphere.

Now the air of a dead planet would be expected to have a composition close to what is called the chemical equilibrium state. That is to say, all possible chemical reactions among the gases would have taken place and the atmosphere would be rather like the exhaust gas escaping from a car or a factory chimney; a mixture from which no more energy could be extracted. A planet that bore life would have a very different atmosphere because living organisms are obliged to use the air as a source of raw materials and as a depository for their waste products. Both of these uses would change the atmosphere away from the chemical equilibrium state. If the observed degree of disequilibrium among the gases of a planetary atmosphere was great, it would suggest the presence of life.

With this in mind, Dian and I were considering what was then known of Mars and Venus. We discovered that both had atmospheres close to equilibrium, like exhaust gases, and both were dominated by the generally unreactive gas carbon dioxide. According to the new theory they both should be dead. To confirm the prediction we looked at the atmosphere of the Earth, the only planet we know to bear life, and found to our delight that it is in a deep state of disequilibrium. Gases that react together such as oxygen and methane coexist, and carbon dioxide is a mere trace gas present at only a few hundred parts per million. Earth's atmosphere is like a dilute form of the energy-rich gas mixture that enters the intake manifold of a car engine before combustion: hydrocarbons and oxygen mixed. An observer on a space craft, even from beyond the Solar System, could confidently have said that, of all the planets, only Earth bore life.

It was at that moment that I glimpsed Gaia and an awesome thought came to me. The Earth's atmosphere was an extraordinary and unstable mixture of gases, yet I knew that it was constant in composition over quite long periods of time. Could it be that life on Earth not only made the atmosphere, but also *regulated* it – keeping it at a constant composition, and at a level favourable for organisms?

Then, I lacked any idea of what the nature of the control system might be, except that the organisms on the Earth's surface must be part of it, and that gaseous composition might be only one of the factors being regulated. I knew, from astrophysicists, that stars increase their heat output as they age and that our Sun has grown in luminosity by 25 per cent since life began. Yet the Earth's temperature has remained comfortable for life for the whole of this 3.8 billion year period. In the long term, could climate also be actively regulated? The notion of a control system involving the whole planet and the life upon it was now firmly established in my mind.

Neither I nor Dian Hitchcock pursued this idea further at the time. We were both struggling with the lesser objective of persuading our life-science colleagues at the Jet Propulsion Laboratory that atmospheric analysis was a valid method of life detection. I must have been very innocent in those days; I now realize that for them to have accepted our idea would have meant admitting that there was almost certainly no life on Mars. Such an admission could have led to the cancellation of the impressive array of life-detection experiments of a more direct kind that were to be carried by the Viking Mission to Mars, and to unemployment among space biologists.

As it was, NASA was very tolerant of our dangerous notion and allowed us to continue working on it. One of our colleagues was the astronomer, Carl Sagan, who was editor of the journal *Icarus*. He disagreed with our arguments for planetary life detection by atmospheric analysis, but good editor and scientist that he was, he invited us to publish our paper in his journal. The first mention of the idea of the Earth as a self-regulating system did not appear until 1968, and then in a brief paper in proceedings of the American Astronautical Society. The name for the self-regulating planet did not come until the early 1970s, and was the gift of the novelist William

Mars

Recognizing a living planet

Imagine yourself aboard a space ship beyond the Solar System but still able to see its planets. The ship has a wide-range spectrometer, with which you can discover the gaseous composition of each planet's atmosphere in turn.

The bar charts illustrate the read-out of the spectrometer for three of the planets: Mars, Venus, and (lowest set) Earth. The abundances of the gases in the charts are in logarithmic units, each step ten times more than the one before. This is to show the trace gases on the same chart as the predominant gases. The ship's on-board computer has also calculated the likely read-out for Earth, were life absent. For convenience, the gases are displayed according to their chemical behaviour, in three separate classes: oxidizing (oxygen and carbon dioxide), reducing (methane and hydrogen), and inert (nitrogen and argon).

The information read-out provides conclusive evidence of life on Earth, and equally conclusive evidence of its absence on Mars and Venus. On Earth, the atmosphere is in a state of great chemical disequilibrium, with reducing and oxidizing gases both present in a highly reactive mixture. Mars and Venus, by contrast, have atmospheres close to the equilibrium state, with oxidizing and inert gases only. In fact, no other planet in the Solar System has oxidizing and reducing gases mixed as on Earth. Nor would the Earth itself if life were absent. Instead it would, as the read-outs show, have an atmosphere with a composition midway between those of Mars and Venus, close to chemical equilibrium, and dominated by the exhaust gas carbon dioxide.

Venus

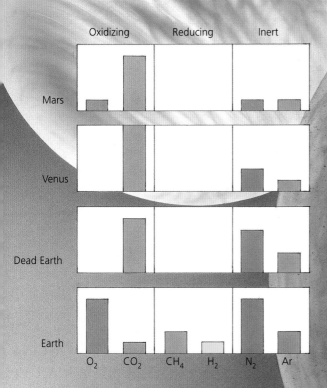

	Oxidizing	Reducing	Inert

Mars

Venus

Dead Earth

Earth

| O_2 | CO_2 | CH_4 | H_2 | N_2 | Ar |

The Earth's improbable atmosphere

The air we breathe is an unlikely cocktail of reactive
oxidizing and reducing gases. Just how unlikely is
demonstrated by the coexistence of oxygen, at 21
per cent, and methane at a fairly constant level of
1.7 ppm. Methane like CO_2 has risen in abundance
as a result of human activities. It was 0.35 ppm in
the glacial period and rose to 0.65 in the early
interglacial. It reacts with oxygen in sunlight to form
carbon dioxide and water. To remain at its constant
level, it must be replenished (by methane-producing
organisms) at a rate of some 500 million tons a year.
It is important to note that these gases stay constant
only for what is the contemporary biosphere. There
was little or no oxygen in the Archean and methane
was probably as abundant as 10 or even 100 ppm.

If life on Earth were suddenly to cease, all the
hundred-plus chemical compounds that make up the
surface, oceans, and atmosphere would react
together until no more change in composition was
possible. The planet would become a hot, waterless,
and inhospitable place with the chemistry of a lifeless
planet; what scientists call the abiotic steady state.

Dead Earth

Earth

Golding, who was then a near neighbour of mine in the village of Bowerchalke in Wiltshire, England. I was, and still am, indebted to him for reminding me that the great system that is the Earth already had a splendid name: Gaia, the Earth goddess, oldest and greatest of the pre-classical Greek pantheon of gods.

If it was easy to see why space scientists were unhappy about a method of planetary life detection that required no more than an Earth-based telescope, it was more difficult, at the time, to see why mainstream science treated the notion with scepticism. It was, after all, the reduction to practice of a science-fictional dream – an instrument that if mounted on a star ship, would let the crew know which, if any, of the planets of a newly discovered solar system bore life.

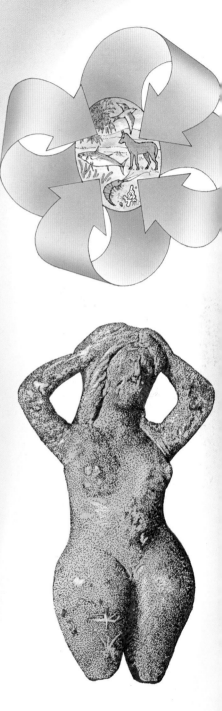

The Gaia hypothesis followed naturally on the idea of life detection by atmospheric analysis. In the early 1970s, I was fortunate to have the distinguished biologist, Lynn Margulis, join with me as a colleague and together we developed the Gaia hypothesis. But when we tried to publish papers on Gaia, we encountered strong resistance. We found it, in fact, impossible to publish Gaia papers in such learned journals as *Science* or *Nature*. There was no lack of interest in Gaia; there was just an apparent determination by peer reviewers (the panels of science arbiters who vet items for acceptance) to deny publication to the idea; indeed, they seemed to regard it as dangerous.

Though publication in fashionable mainstream journals was denied, I was invited to present papers on Gaia at conferences. The invitation often came because the conference organizer saw Gaia as light entertainment, able to provide a break from the more serious, but often tedious, science of the meeting. I accepted because I knew that the proceedings of the conference would be published, and this was our only avenue to publication. Then, in 1988, at the courageous instigation of the climatologist, Stephen Schneider, the American Geophysical Union chose Gaia as the subject of one of its prestigious Chapman conferences, held in honour of the famous aeronomist, Sydney Chapman. Even here, with few exceptions, the papers presented paid lip service only to Gaia; their authors continued to view the Earth through the familiar multifaceted vision of many separated sciences.

Whether or not Gaia is a true view of the Earth is less important than its impressive track record of accurate predictions (see overleaf). Because I view Gaia as a physiological system, I have called the science of Gaia geophysiology. Just as physiology takes a holistic, systems view of ordinary living organisms such as plants, animals and microbes, so geophysiology is the holistic science of large living systems, such as the Earth, in which the organisms are dispersed. Geophysiology is a hard and rigorous science whose study is the properties of these large systems, such as their regulation of climate

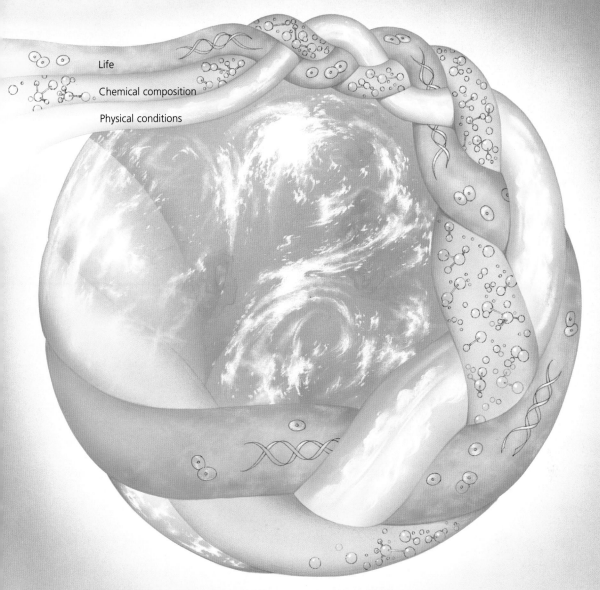

Life

Chemical composition

Physical conditions

The evolution of Gaia

Gaia is the name the ancient Greeks used for the Earth goddess. This goddess, in common with female deities of other early religions, was at once gentle, feminine and nurturing, but also ruthlessly cruel to any who failed to live in harmony with the planet.

Such a name seemed particularly appropriate for the new hypothesis, which took shape in the late 1960s. The Gaia hypothesis was first described in terms of life shaping the environment, rather than the other way around: "Life, or the biosphere, regulates or maintains the climate and the atmospheric composition at an optimum for itself." As understanding of Gaia grew, however, we realized that it was not life or the biosphere that did the regulating but the whole system. We now have Gaia theory, which sees the evolution of organisms as so closely coupled with the evolution of their physical and chemical environment that together they constitute a single evolutionary process, which is self-regulating. Thus the climate, the composition of the rocks, the air, and the oceans, are not just given by geology; they are also the consequences of the presence of life. Through the ceaseless activity of living organisms, conditions on the planet have been kept favourable for life's occupancy for the past 3.8 billion years. Any species that adversely affects the environment, making it less favourable for its progeny, will ultimately be cast out, just as surely as will those weaker members of a species who fail to pass the evolutionary fitness test.

SOME PREDICTIONS FROM GAIA

Whether Gaia is a true picture of the world may be less important than its ability to stimulate the right questions about planetary chemistry, and so open up fertile new areas of research. Its remarkable track record in predicting self-regulatory interactions of biota and the environment is illustrated in the table below. In each case, a question suggested by Gaia theory led to a prediction of a possible link, and subsequent research has either confirmed this prediction as correct, or opened further areas of interest.

PREDICTION AND YEAR	TEST AND RESULT
That Mars was lifeless from atmospheric evidence 1968	Viking Mission 1977 Strong confirmation
That organisms would make compounds that can transfer essential elements from the oceans to the land surfaces 1971	Dimethyl sulphide and methyl iodide both found 1973
That climate may be regulated by the control of carbon dioxide through biologically enhanced rock weathering 1981	Microorganisms greatly increase rock weathering 1989
That climate regulation via cloud density control is linked to algal sulphur gas emissions 1987	Still under test; evidence that oceanic cloud cover geographically matches algal distribution 1990
That oxygen has stayed at 21±5% for the past 350 million years 1973	Plausible models of oxygen regulation now published.
That Archean atmospheric chemistry was dominated by methane 1988	Generally accepted as a working hypothesis
Life greatly enhances the cycling of nutrient elements 1973	Widely confirmed
Life regulates the nutrient balance of the ocean, Redfield 1958.	A pre-Gaian prediction now part of conventional wisdom

Geophysiology: the birth of a new science

Just as physiology is the science of medicine, so geophysiology has arisen as the science of Gaia. Both are systems sciences: physiology is concerned with how living organisms work; geophysiology is concerned with how the living Earth works. Geophysiology ignores the traditional divisions between the Earth and life sciences, which view the evolution of the rocks and the evolution of life as two separate sciences. Instead, geophysiology treats the two processes as a single evolutionary science which, when properly studied, can effectively describe the history of the whole planet.

and temperature. It is also the basis of an empirical practice, planetary medicine. This is not a new scientific discipline; it follows the honourable traditions of scientific thought and experiment. Indeed, it traces back a lineage through such distinguished scientists as G E Hutchinson, Arthur Redfield, Vladimir Vernadsky, Alfred Lotka, and James Hutton. Back to a time when science was whole, and was the vocation of wise and open-hearted natural philosophers. Other scientists who hold similar views of the Earth to mine prefer to call their subject Earth system science, or biogeochemistry. More than anything these names mark the different paths we have taken to reach what may be a common viewpoint. I will stay with geophysiology in this book because of its closer connection to the concept of planetary medicine.

Today, much of science lacks wholeness of vision; it is divided into a bewildering anthology of texts, written in the secluded dialects of reductionist expertises. The numerous tribes of physicists have little knowledge of what they call the "soft" sciences, such as biology. Some physicists even think chemistry soft. Biologists themselves are divided into over 20 different and isolated disciplines. Broad concepts such as life, intelligence, or Gaia, which are comprehensible in a general way to most lay people, are bewildering to scientists. Take the concept of life. Everyone knows what it is, but few if any can explain it. It is not even listed in the *Dictionary of Biology*, compiled by M Abercrombie, C J Hickman, and M L Johnson. If my scientific colleagues are unable even to agree on a definition of life, their objection to Gaia can hardly be rigorously scientific – it may be more often an emotional gut reaction.

The root of the problem is what the philosopher Gilbert Ryle referred to as a category error. He gave an example of such an error by telling of an imaginary alien brought to a university. The alien is shown the classrooms, the administrative offices, the examination halls, and the libraries. The alien meets the students, lecturers, and professors, and then says, thank you for showing me all these, but where is the university?

When we ask "What is life?", we are shown a gamut of living organisms. Mammals first, of course, for toads and frogs seem less alive, and trees and plants less still, and lichens, algae and soil bacteria, hardly alive at all. Much of the instinctive objection to viewing the Earth as a living system comes from our zoocentrism, the tendency to consider ourselves and animals as more alive than other living organisms.

If we ask a group of scientists, "What is life?", they will answer from the restricted viewpoint of their own particular disciplines. A physicist will say that life is a peculiar state of matter that reduces its internal entropy in a flux of free energy, and is characterized by an intricate capacity for self-organization. (Entropy is a measure of the proximity of a system to equilibrium. The lower the entropy,

What is life?

To understand Gaia you must first understand the meaning of the term "life". This simple, everyday concept seems to defy scientific definition. Various dictionaries define life as that property of plants and animals that differentiates them from dead or non-living matter, and that enables them to take in food, extract energy, and grow.

But this definition is woefully inadequate. It ignores the fact that all living systems have boundaries, be they cell walls, membranes, skin or waxy coverings; that they have an ability to maintain a constant internal medium; and that they need a constant flux of energy and matter to maintain their integrity.

The physicist's view

The physicist defines life in terms of a reduction in entropy. By taking in free energy, a living organism is able to decrease its internal entropy, excreting low-grade energy across its boundary.

The neo-Darwinist's view

Life, to a neo-Darwinist, is a property of organic molecular entities that allows them to grow and reproduce. Any errors of reproduction are corrected through the process of natural selection.

The biochemist's view

A biochemist considers an organism to be alive if it is able to utilize free energy (either from sunlight or food) to grow according to its genetic instructions.

the higher the disequilibrium and the greater the information available within the system.) A neo-Darwinist biologist will describe a living organism as one able to reproduce and to correct the errors of reproduction through natural selection among its progeny. To a biochemist, a living organism is one that takes in free energy as sunlight, or chemical potential energy, such as food and oxygen, and uses the energy to grow according to the instructions coded in its genes.

To a geophysiologist, a living organism is a bounded system open to a flux of matter and energy, which is able to keep its internal medium constant in composition, and its physical state intact in a changing environment; it is able to keep in *homeostasis* (see also p. 141). This geophysiologist's description of life includes Gaia. The Earth is bounded on the outside by space with which it radiates energy, sunlight coming in and heat radiation going out. It is bounded on the inside by inner space, the vast volume of plastic hot rock that supports the crust and with which the crust exchanges matter. It is reasonable to consider the Earth as a system in homeostasis. The climate has stayed comfortable for life for 3.8 billion years in spite of an increase in solar output of 25 per cent. The level of oxygen has remained constant for hundreds of millions of years (see p. 113). True, there is not yet proof that the constancy of climate and oxygen has not been kept by some happy accident. But this is no reason to reject the alternative that the Earth is self-regulating like a living organism. To do so in the face of the evidence would be bad science.

Gaia would be a living organism under the physicist's or the biochemist's definitions. The Earth certainly uses solar energy and conducts a kind of metabolism on a planetary scale. It takes in high-grade free energy as sunlight, reduces its entropy within the planetary boundaries, and excretes low-grade energy as infrared to space. It also exchanges chemical materials with the inner space of the Earth's interior.

The strongest objections to Gaia come from the neo-Darwinists – those disciples of Darwin who have combined the great theory of evolution with the discoveries of modern genetics and molecular biology. Gaia cannot reproduce, they say, and therefore cannot evolve in competition with other planets. Therefore it cannot be alive.

It is true that Gaia is not alive like you and me. It has no sense of purpose, it cannot move by its own will, or make love. But then neither can many bacteria.

The category error of the neo-Darwinists is like that of René Descartes (1596–1650), who distinguished humans from all other living things in alone possessing a soul. Both Descartes and neo-Darwinists limit a general property, soul or life, to a specific class of owner. It is possible, of course, to accept this limited view and look on Gaia not as alive, but still as something special, able to regulate

The geophysiologist's view

To a geophysiologist, life is a property of a bounded system that is open to a flux of energy and matter, and that is able to keep its internal conditions constant, despite changing external conditions.

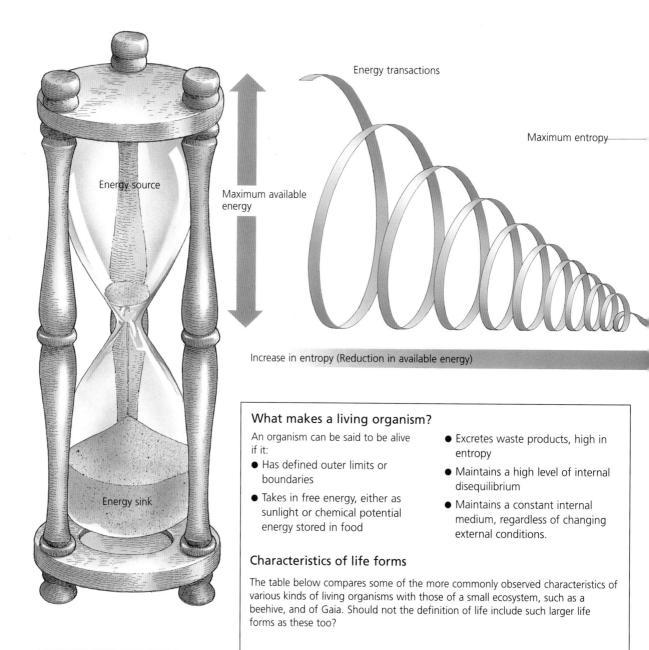

Energy transactions

Maximum entropy

Energy source

Maximum available energy

Energy sink

Increase in entropy (Reduction in available energy)

What makes a living organism?

An organism can be said to be alive if it:

- Has defined outer limits or boundaries
- Takes in free energy, either as sunlight or chemical potential energy stored in food
- Excretes waste products, high in entropy
- Maintains a high level of internal disequilibrium
- Maintains a constant internal medium, regardless of changing external conditions.

Characteristics of life forms

The table below compares some of the more commonly observed characteristics of various kinds of living organisms with those of a small ecosystem, such as a beehive, and of Gaia. Should not the definition of life include such larger life forms as these too?

Entropy

Entropy is a precise physical property like temperature or pressure, popularly but incorrectly associated with disorder alone. It is a measure of proximity to equilibrium. All living things exhibit low entropy – they maintain a high level of internal disequilibrium and abundant information.

CHARACTERISTIC	LIFE FORMS				
	Bacteria	Mammal	Tree	Beehive	Gaia
Reproduction	+	+	+	−	−
Metabolism	+	+	+	+	+
Evolution	+	+	+	+	+
Thermostasis	−	+	−	+	+
Chemostasis	+	+	+	−	+
Self-healing	+	+	+	+	+

Gaia and the second law of thermodynamics

Fundamental to our understanding of Gaia and her place in the universe is an appreciation of the laws of thermodynamics – that branch of physics that deals with time and energy. These laws rule the whole of our universe, and Gaia has had to evolve within their restrictive boundaries.

There are three of these laws and the best way I know of expressing them is:

● You can't win
● You can only ever break even
● You can only break even at absolute zero

There is, however, a more conventional explanation of these rules. For the first law, this states that when one form of energy is converted into another, there is no net loss or gain. Energy, in other words, is always conserved. The second law is harder to swallow. It states that when one form of energy is converted into another, a certain proportion is always lost as heat. From this law we can deduce that the total quality of energy in the universe has direction: it is always running down. Hot objects become cool, but cool objects never become hot. Natural processes are always moving towards thermodynamic equilibrium, that is, a zero-energy state. Maintaining the life of an organism requires a constant input of energy to combat this second law.

Through thinking about Gaia I see our universe as self-organizing and driven by its potential energy which is used as it runs down. The decline of the universe, like the sand trickling through an hourglass, is essential for maintaining the flux of free energy that makes life possible. Indeed, it permits the most amazing self-organization to take place, with us as one of its intricate examples. To be gloomy about the second law is as foolish as expecting to use a flashlight to see in the dark, and have the battery last forever.

itself in a way like a living organism. Something quite different from the dead planets, Mars and Venus. Perhaps neo-Darwinists might accept "quasi-living" as a category for Gaia and for ecosystems, beehives, and other systems that contain organisms and exhibit self-regulation.

I would accept such a categorization but prefer the broad view that includes everything that metabolizes and self-regulates as being alive, so that life is something shared in common by cats and trees, as well as by beehives, forests, coral reefs, and Gaia.

Yet when I talk of a living planet, I am not thinking in an animistic way, of a planet with sentience, or of rocks that can move by their own volition and purpose. I think of anything the Earth may do, such as regulating the climate, as automatic, not through an act of will, and all of it within the strict bounds of science. I respect the views of those with faith who find comfort in a church, and who say their prayers, but acknowledge that they cannot, by logic alone, convince themselves, or others, of their reason for believing in God. Similarly, I respect those who take comfort from the natural world and wish to say their prayers to Gaia.

The redwood tree analogy

One of the most helpful analogies I know of for explaining Gaia was expressed by the American physicist, Jerome Rothstein. I met him at a meeting hosted by the Audubon Society at Amherst in Massachusetts, USA in August 1985. The meeting was called "Is the Earth a Living Organism?". He proposed that the form of life that we all recognize as living, which is closest to Gaia, is the redwood tree, Sequoia gigantea. These noble trees stand in groves on the west coast of America and are spires of lignin and cellulose, 100 metres high and weighing more than 2000 tons. Some of them were a thousand years old at the birth of Christ. The casual tourist to California can observe this date labelled on one of the annular rings of a felled tree; rings that record its 3000 years of life. They are the oldest and the largest things we can see and touch as whole living organisms.

The extraordinary thing about each of these trees is that more than 97 per cent is dead. The wood that is the solid spire is dead, the thick bark around the tree is dead. The only live part is a thin circumferential layer of living cells that enfolds the wood and lies beneath the bark. The leaves, the flowers, and the seeds, of course, are alive but they are a tiny fraction of the mass of the tree.

The redwood tree is like Gaia because the Earth also is made of a vast mass of dead matter, with a thin layer of living organisms encompassed within a protective transparent skin of air. The illustration, overleaf, demonstrates the remarkable resemblance. The wood of the interior and the bark on the outside both were once living tissue but have become just dead wood. The thin circumferential skin of living cells is what keeps the tree alive and growing. In a very similar way the Earth is covered with a surface layer of living organisms, spread thinly over its circumference. Neither the air above nor the rocks beneath are alive, but both have been extensively processed by living organisms, just like the bark and the wood of the tree. The resemblance goes further, for the wood is the structural material of the tree, as are the rocks of the Earth, and both the bark and the atmosphere protect the living matter at the surface from the harsh external environment.

Except for a mere 1 per cent, the gases of the air are wholly the products of surface and ocean living organisms, the 1 per cent exception being the chemically inert noble gases, helium, neon, argon, krypton, and xenon. Even nitrogen, which makes up 78 per cent of the air, comes entirely from organisms, and the other gases, carbon dioxide, oxygen, and methane, are all in rapid exchange with living organisms. Like the bark, which is grown for the protection and sustenance of the living cells of the tree, the air has grown in composition so that it always sustains a favourable climate and a favourable chemical environment for life.

Examining Gaia's medical record

There are other ways too in which Gaia bears more resemblance to a tree than to other life forms. With an animal, virtually the whole of its body is metabolically active during its lifetime. Due to a constant turnover of atoms, little of an animal's past – save its bones and teeth – survives into adulthood. Not so with a tree – or with Gaia. The wood laid down by a tree each year is dead but it preserves a record of the tree's experiences during that year of its life. The tree rings tell us for each year of the tree's life whether it was hot or cold, wet or dry, and what strange elements, or isotopes of elements, were circulating in the environment. The study of tree rings is now a respectable scientific expertise, dendrochronology. As with trees, so it is with Gaia. The rings of sediment laid down around the circumference of the planet every year record the events of that year and become a permanent record of the Earth's history – Gaia's medical record, if you like.

The more recent the evidence, the better its quality. The most impressive evidence comes, as we shall see in Chapter 7, from the layers of ice laid down on the great ice caps of Antarctica and Greenland. Here each year small bubbles of air are trapped in the compressed snow. These bubbles hold intact for tens of thousands of years the air from which they were made. For the first time we have evidence of the changes with time of the atmosphere. These bubbles have revealed how different the air was in the last ice age, how suddenly it changed at the end of the glaciation some 10,000 years ago, and what have been the effects of people on the planet. When I talk of the constancy of our regulated atmosphere I am, like a

How does Gaia resemble a redwood tree?

If you find it hard to believe that anything as large and seemingly inanimate as the Earth is alive in any sense of the word, then it may help to compare the planet with a giant redwood tree. The tree is undoubtedly alive, yet more than 97 per cent of it is composed of dead wood. The thin circumferential skin of living cells (known as the cambium) just beneath the bark is what keeps the tree alive and growing.

In a similar way the Earth has also a "cambium" composed of the surface layer of living organisms spread thinly over its circumference. The bark and the atmosphere both protect the living matter at the surface. All the gases of the air – nitrogen, oxygen, carbon dioxide, and methane – are the direct products of living organisms, except for the 1 per cent contribution made by the noble gases, such as argon.

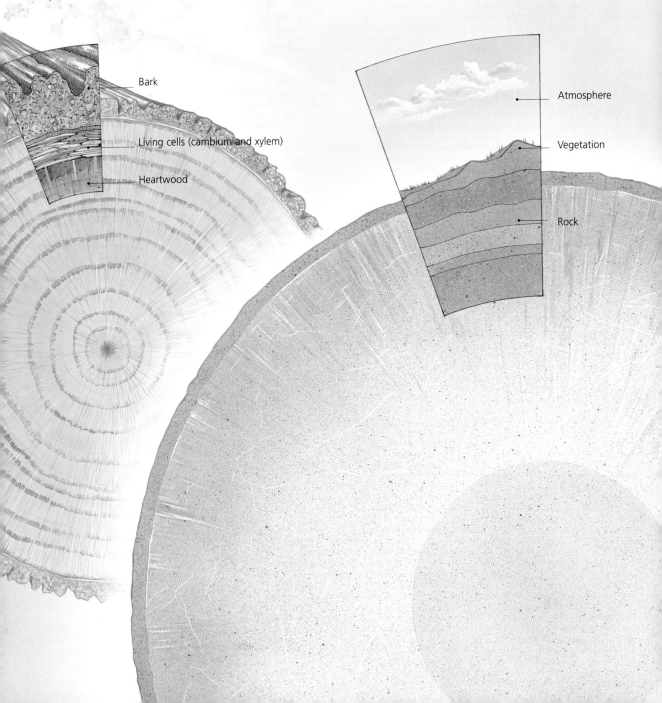

Bark

Living cells (cambium and xylem)

Heartwood

Atmosphere

Vegetation

Rock

physician, thinking of constancy within permissible bounds. Our normal temperature is within a degree of 37°C but when we are feverish it can reach 40°C. The Earth's temperature covers a similar range from about 10 to 14°C between its cold and warm periods. Both represent changes within acceptable bounds. The same type of variation within permissible bounds applies to atmospheric gas abundances.

But before I delve further into the fascinating life history of our planet, let's pause to take a fresh look at the anatomy of this living entity today. The next chapter describes this superorganism called Gaia, and asks what are its tissues and its organs, and what roles do they play in sustaining the living Earth? It is a chapter designed to re-orient you, to help you see the Earth you thought you knew through the perspective of Gaia. For this snapshot view, we shall have to touch on many subjects explored in more detail later in this book. And to gain a clear understanding of Gaia's anatomy, we shall combine the holistic, "top-down" view of the living planet, with a reductionist, "bottom-up" analysis of Gaia's working parts.

CHAPTER TWO

Anatomy

The very word "anatomy" means dissection. It carries with it the vision of cadavers, lying ignominiously pickled in formaldehyde in a medical school, waiting to be cut open so that the fledgling medical students will know which part of us is which and does what. It is the essence of reductionism so to examine by dissection what once was a living organism.

So how can we analyse the Earth, while it is still alive, anatomically? The answer lies in the slow but happy movement of science away from vivisection and dissection towards non-invasive analysis. We no longer have to cut up a human being to reveal the structure. We can see from outside the body how it is arranged using radiations that can penetrate without significant harm. Roentgen's application of his newly discovered X-rays to view the skeleton from outside was the first step in this direction for human medicine, and soon developed into the routine procedures we all know so well. The safer methods of ultrasonic and magnetic resonance imaging came later.

We do the same with the Earth. There are radiations that let us see the structure of the oceans, the atmosphere, and the rocks. Visual, infrared, ultraviolet, and radio-frequency radiations have all been used to look down at the Earth from above and up from the surface beneath. By this means the layered structure of the atmosphere was revealed, with the ionized layers high up and below them the ozone layer of the stratosphere (see p. 41). Using ultrasound we can probe the depths of the oceans, and with the noise from earthquakes and explosions, the structure of the solid Earth itself.

In this book I extol the virtue of the top-down, holistic approach and often chide my colleagues in science for being too reductionist, too narrowly focused in their view. But in this chapter about the anatomy of the Earth, I have to acknowledge the value of reduction in science, and recognize, as would a sensitive medical student, that analysis by anatomy is a necessary step in understanding a living body.

Not all things reductionist are bad, nor are all things holistic good. The reductionist, bottom-up view can be needed just as much as the holistic. One of the great rewards of science is that sudden flash of understanding that comes when holism and reduction meet. A meeting as exciting and joyful as that of tunnellers in the heart of a mountain or beneath the sea, who meet their fellows digging from the other side. Such a meeting of minds took place in the mid-20th century, when nuclear physicists and cosmologists realized that they shared a common view of the universe.

With this in mind, let's take a snapshot view of the whole planet, as seen from space, and then go on to look at the structure and layout of its main compartments: the atmosphere, the waters, and the rocks. Finally we shall examine the great ecosystems themselves, before putting all the pieces together in "The ensemble" (p. 56). Although this chapter is concerned only with the parts, the components of Gaia, we still need to keep in mind the holistic view: how the parts and their functions all relate to each other.

Gaia: a holistic view

Seen from space, the Earth looks entirely different from the other terrestrial planets, Venus and Mars. Abundant water and oxygen give the Earth a characteristic dappled, blue and white appearance. Mars and Venus are dull and dun by comparison. They stand to the Earth as the sands of the desert to a freshly blooming flower. Viewed in the radiations of other wavelengths that our eyes do not see, the difference is even greater. The Earth's reflection of infrared radiation is intricately modulated by the vibrations of molecules such as methane, nitrous oxide, methyl chloride, and ozone in the atmosphere. In these wavelengths, the air itself is multicoloured. In the ultraviolet, reflections reveal the presence of oxygen. The prolonged co-existence of oxygen and reactive gases such as methane is something quite impossible in the atmosphere of a dead planet. It is unique to the Earth. The view from space reveals other differences too. The Earth's total output of infrared radiation (that is, its total loss of heat) is less than would be expected for a planet in its position in the Solar System, or were it bare rock like the Moon, and much less than if it had an atmosphere that retained just the right amount of heat to sustain a climate comfortable for living organisms.

It was this top-down view of the Earth from space that led to the realization that the Earth's unique, yet unstable and reactive atmosphere, would need some means for its long-term regulation – a means such as Gaia.

Dissecting the parts of Gaia

The idea of dissection applied to the Earth is almost self-contradictory. It implies a holistic view of the planet – seeing it as an organism – but uses an approach that is reductionist. The problem with reductionism lies with the belief that the method of examining systems by taking them apart is all that is needed. Reductionists are certain that there is

Portrait of Gaia

Seen in all its shining beauty against the deep darkness of space, the Earth looks very much alive. This impression of life is real. Only a planet with abundant life, and able to retain its water and regulate its unique atmosphere and climate, could appear so different from its sister planets, Mars and Venus, both of which are dead. Of course the Earth is not alive like an animal, able to reproduce itself and have its progeny evolve in competition with other animals. It is a superorganism, alive like the great ecosystems or some giant tree, the largest life form we yet know. I think it wrong of science to deny the status of life to such entities, a state of life intermediate between inanimate matter and a sentient organism, yet greater and longer lived than most organisms.

A spherical shell

Anatomically Gaia is a spherical shell with its most active part at the Earth's surface, but existing also above it in the atmosphere, and below it in the deep oceans and crustal rocks. In the 3.8 billion years of life, Gaia has conducted energy transactions with space, receiving high potential visible and ultraviolet radiation from the Sun and discarding low potential infrared to space. In a similar way Gaia has traded with the inner space of the Earth's interior, exchanging gases and minerals. So small is the layer of Gaia and so vast is space and the interior that we can take them both to be infinite sources and sinks unaffected by what Gaia does.

Infrared radiation

Solar energy

Ultraviolet radiation

Facts of the Earth

The table below lists the Earth's "vital statistics". 6.371×10^6 means $6.371 \times 1,000,000$ (6 zeros).

Solar constant	1.375 kW/m^2
Earth mass	$5.976 \times 10^{24} \text{kg}$
Radius	$6.371 \times 10^6 \text{m}$
Surface area	$5.101 \times 10^{14} \text{m}^2$
Land area	$1.481 \times 10^{14} \text{m}^2$
Ocean area	$3.620 \times 10^{14} \text{m}^2$
Mean land elevation	$8.40 \times 10^2 \text{m}$
Mean sea depth	$3.730 \times 10^3 \text{m}$
Ocean volume	$1.350 \times 10^{18} \text{m}^3$
Ocean mass	$1.384 \times 10^{21} \text{kg}$
Atmospheric mass	$5.137 \times 10^{18} \text{kg}$

nothing in the whole system that cannot be predicted from a knowledge of the parts. I acknowledge that this simple faith in determinism has inspired most of the great scientific discoveries so far, but as understanding grows, we are more and more aware that the universe we inhabit is both self-organizing and in many ways unpredictable. To understand it and its most complex entities – living systems – reduction alone is not enough.

If we are to examine the Earth anatomically and ask what are its tissues and its organs, we have to regard it as an entity with some of the properties of a living organism. Even if you reject the idea of a living Earth entirely, you will surely concede it to be helpful in understanding such a complex entity to reduce it to its parts.

With living organisms the intensity of life varies from part to part. Your hair, your nails, and the outer layers of your teeth contain no living cells, yet they are undeniably a part of you. So it is with the atmosphere, the oceans and the crustal rocks of the Earth: they are parts of the organism in which life is thinly dispersed. Birds and insects in flight or people in airplanes hardly affect the bulk composition and properties of the atmosphere. The oceans are more populated and the soil even more so, but still the atmosphere, the sea, and the rocks are usually seen as lifeless in themselves. I, however, see them as essential parts of a larger organism.

Like the feathers of a bird or the fur of a cat, the atmosphere keeps the surface warm. Without its natural greenhouse gases, the surface would be frigid, at a mean temperature of –19°C. The atmosphere also, like skin and fur, shields the delicate living cells of the surface against exposure to solar radiation. The waters of the Earth, as James Hutton saw long ago, are like the circulation system of an animal. Their ceaseless motions (together with the blowing of the wind) transfer essential nutrient elements from one part to another and carry away the waste products of metabolism. The rocks themselves are like our bones, both a solid strong support and a reservoir of mineral nutrients. The rocks are not static; continual movements of the plates and volcanic activity transfer solid and gaseous material from the magma to the atmosphere and oceans. Plate movements also draw down, and mix in with the magma, sediments that originated in the surface environment.

So let us first consider the anatomy of these parts before considering that of the great ecosystems, which are manifestly alive. The table on page 37 gives the Earth's "vital statistics". You can see, for example, that the oceans cover about 70 per cent of the Earth's surface.

The atmosphere

The atmosphere is the least massive part of the physical Earth, weighing not much more than a billion megatons; the ocean is a thousand times more massive, and the Earth itself a million times more. Yet without the atmosphere life could never have begun, nor have existed at any time.

So important is the atmosphere as a tissue of Gaia that no single map or diagram can illustrate all of its complex interactions. The illustrations here try to fill the picture by considering in turn its structure, its circulation, a tissue sample, and an analysis of its composition and possible life functions.

The atmosphere has a structure of concentric shells like an onion. At the top, merging into space, is the thermosphere – a near vacuum with only a few hundred atoms and molecules per millilitre. It is quite hot (several hundred degrees Celsius) but so tenuous that the heat has no effect on the Earth's climate. Yet the thermosphere is vitally important to life on Earth, since conditions here determine how much hydrogen is lost to space. The hotter the thermosphere, the more hydrogen is lost – and in the long run the more water is lost, since water is the only source of hydrogen.

Below the thermosphere lies the mesosphere, where the fierce energy of unfiltered sunlight breaks down the molecules of the air: carbon dioxide splits into oxygen atoms and carbon monoxide; water into hydrogen atoms and hydroxyl radicals; oxygen into oxygen atoms.

Next comes the stratosphere, so called because it shows little vertical mixing. Its chemistry is dominated by the creation and accumulation of ozone, to form the protective ozone layer at around 30 kilometres above the surface, and by the destruction of methane and nitrous oxide and the chloroflourocarbons (CFCs) rising from below. (Some of the first two, and all of the CFCs, are the result of human activity.) Ozone is both made and destroyed by the energy of sunlight, but some of its destruction is catalyzed by chlorine and bromine from the decomposing CFCs and halons.

In the stratosphere are layers of aerosols, especially droplets of sulphuric and nitric acids. These can become quite dense after large volcanic eruptions, causing a lowering of Earth's surface temperature by reflecting sunlight back to space. You would see them as a milkiness in the sky by day, or as brilliant fiery red and deep violet sunsets at evening. In recent years, clouds of ice crystals have appeared in the stratosphere, possibly due to increasing water vapour, itself a by-product of methane oxidation in the stratosphere. (Methane has more than doubled in abundance since humans began farming.) The presence of acids and ice crystals in the stratosphere is connected with the formation of the "ozone hole" over Antarctica.

The bottom layer of the atmosphere is the troposphere. This is where we live, and experience climate and weather. The air is vigorously mixed for much of the time, and most of the clouds and aerosols are here. Short-lived gases such as ozone, dimethyl sulphide, and methyl iodide (see p. 120), and many of the products of pollution, are concentrated near the surface. But gases with lifetimes in the atmosphere of a year or more – oxygen, carbon dioxide, and methane – are more uniformly spread through the troposphere; contrary to popular belief, heavier gases such as carbon dioxide have no tendency to concentrate lower down due to gravity. Water vapour concentration is the most variable, ranging from several per cent of the air at the surface to one part per million at the top of the troposphere in the tropics.

Energy flows

Solar energy is by far the most important source of energy to the planet. It heats the atmosphere and oceans, particularly in the tropics, driving the global circulation of these two planetary systems. The composition of the atmosphere is critical in warming the Earth's surface: certain trace gases (carbon dioxide, methane, and water vapour), plus the presence of clouds, aerosols, and particulate matter, play a major role in absorbing long wave infrared radiation. Without this warming effect, the temperature at the Earth's surface would be 33°C colder than it is now.

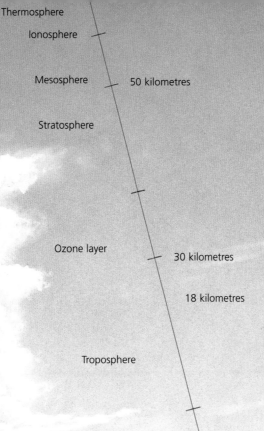

Thermosphere

Ionosphere

Mesosphere — 50 kilometres

Stratosphere

Ozone layer — 30 kilometres

18 kilometres

Troposphere

Layers in the atmosphere

The illustration (left) shows the layered structure of the Earth's atmosphere. From 90 kilometres above the surface, merging upward into space, is the **thermosphere**. The hot layer also contains much of the **ionosphere**, since some of its molecules and atoms are split into free electrons and positive ions by the Sun's ionizing radiations. From 90 down to 50 kilometres lies the **mesosphere**, a region of fierce unfiltered sunlight, ranging in temperature from 0°C at the bottom to −100°C at the top. From 50 down to 18 kilometres the **stratosphere** (so called because there is little vertical mixing between the layers) has temperatures from 0°C at the top to −70°C at the bottom. This region contains the **ozone layer**, with its maximum abundance about 30 kilometres above the surface. The **troposphere** ranges from the surface up to 7 kilometres at the poles and 28 kilometres in the tropics, its temperature ranging from an average of 14°C to −70°C where it meets the stratosphere.

The moving air

The troposphere, where we live and where the biota actively exchange gases with the air, is vigorously stirred by wind and weather, circulating moisture, nutrients, and wastes around the globe, and between oceans and land. The mixing is in three dimensions – the troposphere cools as it extends upward, so it is easy for warm air to rise and clouds to form. But there is little mixing between the north and south hemispheres: air is warmed more at the equator, and this creates self-circulating wind currents either side with the windless "doldrums" between. Such global wind patterns can be mapped but weather is unpredictable – and we increasingly perceive that organisms may play a part. Both ocean algae and forests, for instance, are involved in cloud formation, with its associated wind and rainfall patterns.

Doldrums

What's in a sample of air?

A tissue sample of the air, taken near the surface, shows an energetic and unstable mix of gases: oxygen at 21 per cent co-existing with methane, though they are highly reactive; low carbon dioxide, implying active regulation of this gas; and many trace gases of biochemical origin, the aerial messengers, scavengers, and carriers of Gaia (analogous to those within our own body chemistry). Living organisms are thinly spread travellers (though myriads of spores, seeds, and microbes use the air for dispersion). Near the ground, leaves extend the gaseous exchange boundary, but the air itself is a biological product, an extension of Gaia's metabolism, a result of active exchange of gases with living organisms. The air is protective skin, a warming blanket, an exchange and water circulation medium.

Some chemically reactive gases of the air

Gas	Abundance (%)	Flux (Mt)	Extent of disequil-ibrium	Feedbacks
Nitrogen	79	300	10^{10}	Atmospheric pressure Fire control Ocean nitrate sink
Oxygen	21	100,000	None Taken as reference	Reference level for energy
Carbon dioxide	0.03	140,000	10^3	Photosynthesis Climate
Methane	10^{-4}	500	Infinite	Ozone and nitrogen
Nitrous oxide	10^{-5}	30	10^{13}	Oxygen and carbon Ventilation of anoxic zone
Ammonia	10^{-6}	300	Infinite	pH and nitrogen
Dimethyl sulphide	10^{-8}	70	Infinite	Sulphur cycle Salt stress Climate
Methyl chloride	10^{-7}	10	Infinite	Stratospheric ozone
Methyl iodide	10^{-10}	1	Infinite	Iodine cycle Climate

Gases of the air and their possible system functions

The table (left) shows the abundance and flux (rate of annual flow through the atmosphere) of some of the gases of the air, and the extent to which this composition seems to violate the ordinary rules of equilibrium chemistry. (The word "infinite" in this context means beyond the limits of computation.) The right-hand column of the table suggests some possible feedback control functions that these gases might have in the system of life and its environment, according to Gaia science. These are explored in more detail in later chapters, particularly Chapters 6 and 7.

Clouds

The destructive and unpredictable power of the hurricane (above) reminds us of how little we know about weather and climate. We are astonishingly ignorant about clouds. The common assumption that they just form from water vapour in the air is wrong. Clouds form where there are condensation nuclei – tiny drops or particles of hygroscopic material, often the oxidation products of sulphur compounds. Gaia science has shown that marine algae produce abundant sulphur gases, so raising the awesome possibility that the cloud cover of the Earth might be determined by life (see p. 127). Clouds affect global climate. Low-lying clouds such as marine stratus tend to cool by reflecting sunlight back out to space; high thin clouds, such as cirrus, warm by reflecting heat back to the ground. The net cloud effect is probably cooling, but there is much to learn yet.

The eye of the hurricane

A computer-generated view of Hurricane Allen, Gulf of Mexico, 1980.

The waters of the earth

Without water there can be no life; and (as Chapter 4 relates) without life there would be no water. Life enabled the Earth to hold onto its oceans, and they now dominate its surface. As remarkable as the presence of water is the fact that ocean salinity appears never to have exceeded the critical limit for life. This suggests that it may be regulated by long-term biological and tectonic processes (see p. 48), which prevent the oceans' salt content rising further.

The euphotic zones of the oceans where sunlight is plentiful are the principal seats of activity. We are beginning to recognize the profound significance of these ecosystems to the regulation of climate (p. 50). They may be key organs of Gaia. The surface layers, much warmer than those below, are stratified and separated from them just as is the stratosphere from the troposphere. The 4500 metres of ocean below them is by comparison almost lifeless. However, a more vigorous suite of organisms lives near the sea-floor spreading zones, where the hot basalt rock wells up and the water is rich in iron and sulphides.

Terrestrial inputs
DOC

Ocean currents

No two-dimensional, static map can show the three-dimensional ceaseless flow and mixing of the ocean, like the circulatory system of an animal. In places the oceans are nearly 10 kilometres deep; the water is cold and pressures close to 1 ton per square centimetre. Deep and cold ocean currents travel thousands of kilometres from polar regions before upwelling to the surface, bringing nutrients such as phosphorus and nitrogen, which are scarce there. Surface currents spread heat from warmer tropical waters to colder regions.

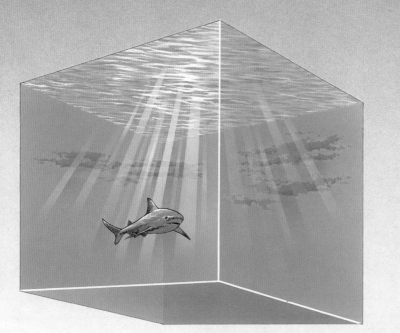

What's in a sample of ocean?

A tissue sample from the surface waters of the ocean reveals many signs of Gaia. It would not seem particularly rich in life to the eye, but a microscope would show blooms of photosynthetic algae grazed by tiny zooplankton, which in turn provide food for larger marine organisms. In their intricate shells, the microflora segregate calcium and silicon, which later rain to the sea bed to form sediments – which in turn may be linked to the movements of the Earth's plates. Salinity is within the tolerance levels of life, and dissolved carbon dioxide is present, pumped down from the air. Also present are trace gases released by the algae, such as dimethyl sulphide, which we now suspect to have a role in cloud formation and climate.

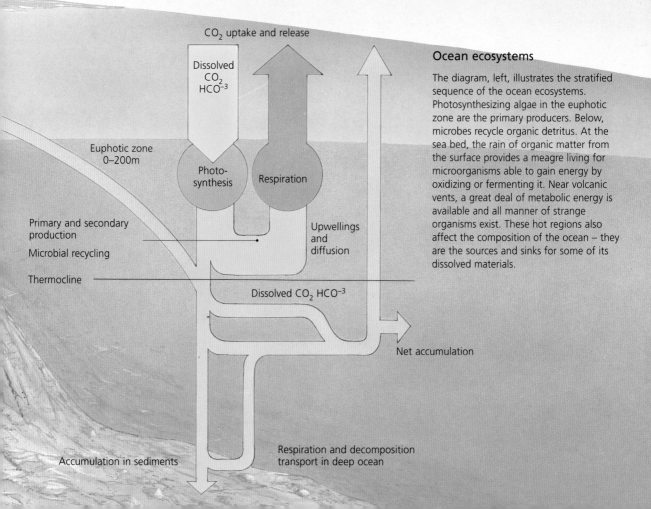

CO₂ uptake and release

Dissolved
CO₂
HCO⁻³

Euphotic zone
0–200m

Photo-synthesis

Respiration

Primary and secondary production

Microbial recycling

Thermocline

Upwellings and diffusion

Dissolved CO₂ HCO⁻³

Net accumulation

Accumulation in sediments

Respiration and decomposition transport in deep ocean

Ocean ecosystems

The diagram, left, illustrates the stratified sequence of the ocean ecosystems. Photosynthesizing algae in the euphotic zone are the primary producers. Below, microbes recycle organic detritus. At the sea bed, the rain of organic matter from the surface provides a meagre living for microorganisms able to gain energy by oxidizing or fermenting it. Near volcanic vents, a great deal of metabolic energy is available and all manner of strange organisms exist. These hot regions also affect the composition of the ocean – they are the sources and sinks for some of its dissolved materials.

Solute	Abundance	Significance
Positive ions		
Sodium	0.468	The principal ions
Magnesium	0.053	of sea salt
Calcium	0.010	
Potassium	0.010	
Negative ions		
Chloride	0.545	
Sulphate	0.028	
Bicarbonate	0.0025	
Bromide	0.00084	
Charge neutral salt		
Dimethyl sulphonio propionate	10^{-10}	Precursor of DMS
Toxic elements		
Mercury	2×10^{-10}	Neutralized by
Lead	1×10^{-10}	organisms
Cadmium	9×10^{-10}	
Nutritious elements		
Zinc	7.5×10^{-8}	Collected and
Iron	4×10^{-8}	concentrated by
Cobalt	9×10^{-10}	organisms
Phosphorus	1.9×10^{-6}	
Nitrogen	3.6×10^{-5}	
Iodine	4.7×10^{-7}	

Solutes and ocean systems

The table (left) shows some of the chemicals found in solution in the ocean, and their significance for marine ecosystems and Gaia. The curiously named dimethyl sulphonio propionate is used in the biochemistry of some marine algae, apparently to relieve salt stress. Its production leads to release of the gas dimethyl sulphide (DMS), which is linked to the global sulphur cycle and possibly to cloud formation and climate control (see Chapters 6 and 7).

NB Abundance expressed in molar units.

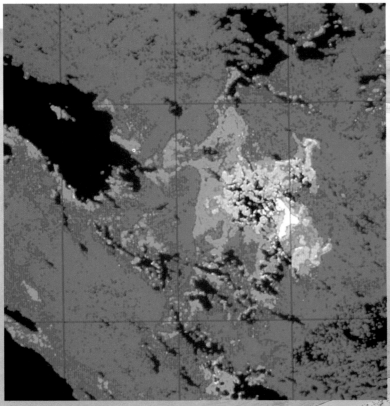

Open ocean

Coccolithophore bloom in the North Atlantic. The photograph is computer-enhanced by colour to show the variations in algal density. The red/orange colours show the greater densities, while the green/blue/purple indicate the lower densities.

Gaia's plastic skin

Three times as deep on the continents as under the oceans, the Earth's crust is not fixed but formed of huge islands or "plates" floating slowly over a basalt sea – the magma of the mantle. More rigid at the top, but plastic at the asthenosphere lower down, under the oceans, the crustal plates show great rift lines, through which hot magma wells up; and on either side of these the sea floor spreads away, pushing against the continents causing folding and mountain building, and dragging the sediments of life down under them, to merge again with the mantle. It is tempting to link these tectonic processes with Gaia (that is with the effect of limestone sediments on the crustal movement (see Chapter 6) and the link with ocean salinity).

The rocks

The domain of Gaia extends through the Earth's crustal rocks, all parts of which have been profoundly changed by the three and a half billion years of life. Surface weathering in the presence of microorganisms has converted the original rock rubble to soils, and washed the early rocks away as silts and solutes to the sea. On the ocean floor, the rain of detritus from eons of life has built up great layers of sediment for burial, and later uplift as sedimentary rocks. The white cliffs of Southern England (see p. 3) are a dramatic example of this life-formed landscape. All the sedimentary rocks have been shaped by living organisms and surprisingly, some less likely features, such as microbially collected mineral deposits of uranium or gold (see pp. 86–7). Volcanism constantly replenishes the surface with molten material and gases from the hot and radioactive interior. So vast is the interior of the thin film of life and the crustal rocks, that it can be taken as an infinite sink and source of materials.

What's in a sample of soil?

A sample of the earth and rock surface beneath reveals the hidden powerhouse of Gaia, the bacterial ecosystems. No cubic centimetre of the soil and sediment beneath is without its billions of microscopic organisms: the topsoil has its photosynthesizing bacteria, and nitrogen-fixing microbes often attached to the roots of plants; its fungi, stones, moulds, and teeming invertebrate life. The anaerobic dwellers of the sediment are ceaselessly active in breaking down organic matter, to release again the nutrients of life. The air in the soil is rich in carbon dioxide pumped down by life; dissolved in water near the rock surface, this causes rock weathering, speeded by microorganisms. Without life there would be no soil, but only regolith, the rock rubble of dead planets.

Global plate tectonics

The half-globe (right) shows some of the principal plate boundaries of the Earth's crust. The long mid-Atlantic ridge is clearly visible at the centre. It is marked by active volcanism and sea-floor spreading along its length.

The active Earth

Volcanic activity constantly replenishes the Earth's surface with molten material and gases. The photograph above shows the eruption of Mauna Loa, Big Island, Hawaii, in Spring 1984.

The great ecosystems

Conventional science defines an ecosystem as a stable, self-perpetuating system, composed of a community of living organisms occupying a non-living environment. According to this view, organisms do not alter their environment, they merely adapt to it. The Gaian view of an ecosystem, however, sees the two components of the system, the living and the non-living, as two tightly coupled interactive forces, each one shaping and affecting the other.

The oldest and greatest ecosystems are the euphotic zones of the oceans, where microscopic algae live; the anaerobic sediments, where simple bacteria biodegrade organic debris, and the aerobic bacterial life of the soils. Most recently arrived on the Earth are the ecosystems of the forests, both temperate and tropical, and the human farms and farmed forests. These are what we usually see as ecosystems, but even here the bacterial infrastructure is crucial.

The natural ecosystems make up the system Gaia. In some ways they correspond to our organs, the liver, the blood, the skin, and the lungs. Each has a partial independence, vital to the system, but is unable to exist except as part of that system. Let's take a look at the way these ecosystems might serve within the whole system Gaia.

The ocean algal ecosystem in the northern and southern arctic and temperate oceans is active chemically in pumping down carbon dioxide from the air, and is also a source of sulphur, selenium, and iodine gases as well as of hydrocarbons, nitrogen oxides, and to a small extent, methane, in the air. The fall of tiny shells from the dead algae (coccolithophores) is responsible for calcium carbonate and silica deposition in the sediments of the ocean floor, and in these shells the algae also segregate toxic elements. We now have little doubt that a significant climatic role is attributable to this ecosystem, whose presence affects both carbon dioxide and clouds in the atmosphere (see p. 147).

In the tropical oceans, life is much sparser. Nonetheless these regions play a significant climatic role for the whole planet, for it is here that solar radiation is most absorbed. There is some evidence that algal growth in these regions occurs in such a way that the water is kept transparent, so that the Sun's radiation penetrates deeply (see also p. 44). We still know little about this vast ecosystem and its role in planetary climate regulation. The coastal waters and those above the continental shelves are the most fertile of the ocean ecosystems. They seem to play an important part in sedimentary processes.

On land, the great ecosystems of the anaerobic sediments and the soil play fundamental roles in the chemistry of Gaia – from methane generation to rock weathering and the regulation of the carbon cycle (see Chapter 6). Younger than these are the forests. The northern and southern temperate forests cover about 10 per cent of the land area. Through their dark colour and capacity to shed snow, conifer forests may lessen the length of winter in near-arctic regions.

Ecosystems: organs of Gaia

As the illustration suggests, ecosystems are spread across the surface of the Earth, from the cold tundra and mountain tops, through the temperate and tropical forests, grasslands and deserts, swamps and coastal waters to the deep oceans. What the diagram cannot show is the great bacterial ecosystems that sustain the rest: the photosynthesizers and consumers always at the surface of soil and sea, the anaerobic fermenters beneath. Ecosystems comprise living organisms and their environment, tightly coupled as single evolving domains. They can be seen as superorganisms, having some of the characteristics of living entities: self-regulation, homeostasis, metabolism. They are also organs of Gaia – each with a distinct identity, yet ultimately interlinked with all the other ecosystems; and with a vital role in the whole organism.

CASE HISTORY A Case of Structural Deformity

The case history I will use to illustrate this chapter on anatomy concerns the Permian period of the Earth's recent history. It is about a malaise that was similar to a structural deformity at birth in a human, like a badly curved spine. In the Permian all of the continents were stuck together in a large lump called "Pangaea", centred near the equator, leaving the polar regions as oceans. Gaia was able to regulate climate within the range comfortable for life but the distribution of temperature and rainfall was not as good as it might have been. Gaia could be said to have been in a state of chronic disease.

The Permian was from 280 to 225 million years ago. The congregation of continents huddled together are thought to have been a vast island like, but much larger than, Australia now. There were shallow seas poorly connected to the oceans, so that they were continuously drying out and depositing salt.

Some of these vast salt beds now lie under northern Europe. I once visited the cathedral-like caverns of the salt mines of Cheshire, England, from which is gathered the rock salt scattered on British roads during icy spells in the winter. The salt layers, now hundreds of metres below the surface, stretch across Britain and on into the North Sea, into the Netherlands, Germany, Poland and the USSR.

The Permian ended when the continents began to separate and drift apart. Smaller and more manageable land masses meant better climate and distribution of rainfall. Reptiles dominated the land, sea, and air, and the shallow seas were rich in marine life such as ammonites and clams. By the end of the Jurassic period, some 140 million years ago, the world was ready to start the luxuriant abundance of the Cretaceous.

Pangaea

Some 200 million years ago, the continents were joined together in a single land mass – Pangaea – surrounded by one ocean, known as Panthalassa. Earth scientists have been able to reconstruct the form of this ancient supercontinent by fitting together the outlines of present day land masses (see diagram, right). With all its exposed land locked into one region of the globe as Pangaea, the Earth's climate regulation would have been at below maximum efficiency. As we have seen, life itself may have been involved in the process of continental drifting that then began. As a result Pangaea broke up into two continents – Laurasia and Gondwanaland – separated by the Tethys Sea.

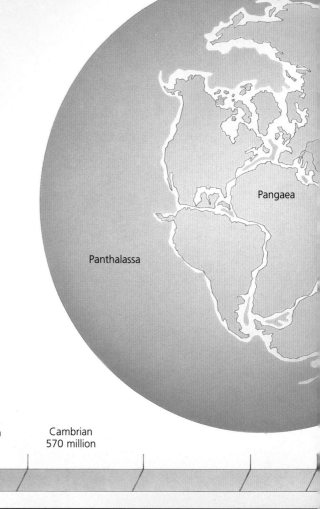

The time dimension

In terms of the history of the planet, the formation of Pangaea and its subsequent disruption into smaller land masses has been a relatively recent phenomenon, occurring about halfway through the Phanerozoic era. The proliferation of flowering plants, birds, and mammals didn't start until the Cretaceous period, some 140 million years ago, at a time when the continents had already separated.

Pre-Cambrian
4.6 billion

Cambrian
570 million

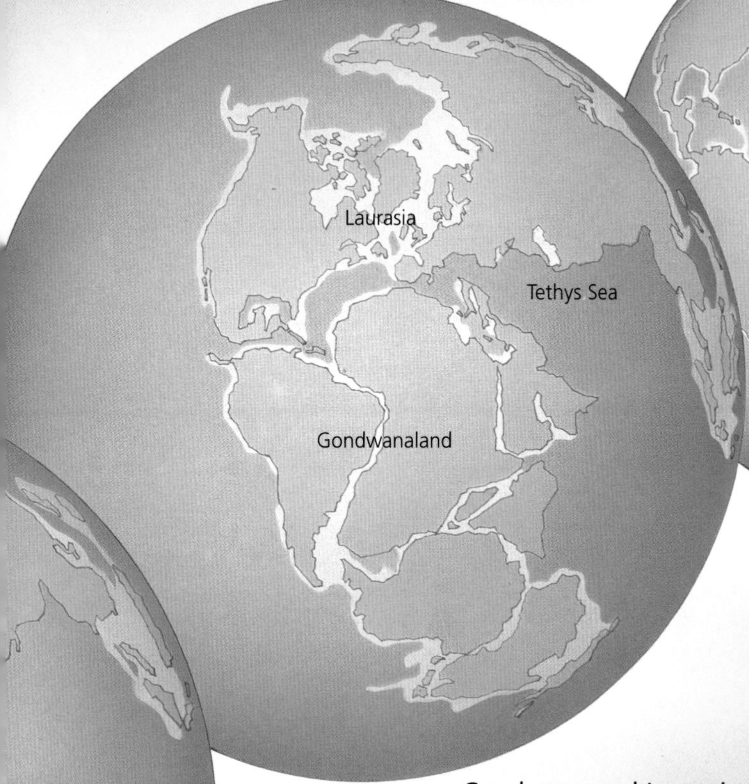

Laurasia

Tethys Sea

Gondwanaland

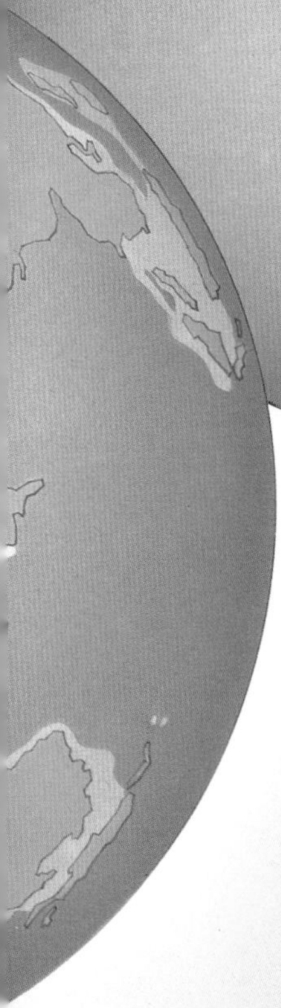

Present-day Earth

The present familiar outline of the major continents is the consequence of further rifting. Australia broke away from Antarctica, and North America from Eurasia. This process of continental drifting has probably been going on for 3.5 billion years and continues even to this day – possibly still driven on, as at first, by the limestone deposition of ocean algae around the continental margins. Eventually, when the Earth's internal heat production becomes too small to sustain the energies of tectonic movement of the plates, then its surface may become stable, as has already happened on Mars – unless life again finds a solution.

Gondwana and Laurasia

Movement of the land masses didn't stop at the formation of two continents. Great rifts resulting from the movement of the Earth's tectonic plates caused Africa and South America to split off from the rest of Gondwana and move north-westward. India also parted company with Gondwana and drifted northward to impinge violently on the Asian part of the continent Laurasia. The Himalayas are the result of this massive buckling of the Earth's surface.

Permian
280 million

Cretaceous
140 million

Present

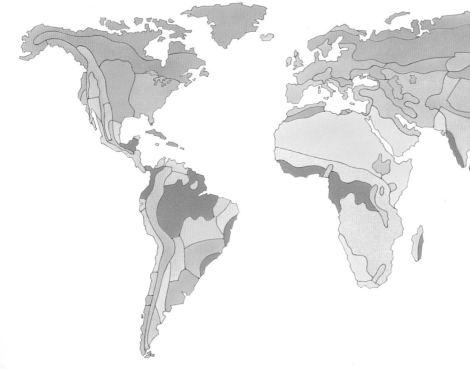

The rainforest

Sunlight shining through the canopy of the tropical rainforest (left). You can see the water vapour forming in the air as the trees evapotranspire.

Ecosystems of the Earth

The map below shows the distribution of 14 major "biomes" on the land surface. (It is based on the work of Miklos Udvardy and IUCN sources.) Like many other such maps it can tell us nothing about the great microbial ecosystems of soil and sediment. But it does show, according to conventional scientific wisdom, an accurate picture of the land ecosystems.

Tropical humid forest

Temperate forest

Temperate needle-leaf forests

Tropical dry forest/woodland

Temperate broadleaf forest

Evergreen sclerophyllous forest

Warm deserts

Cold winter deserts

Tundra

Savannah

Temperate grassland

Mountain systems

Island systems

Lakes

This suggestion has now received support from Richard Betts and Peter Cox, scientists at the UK Hadley Centre for Climate Research. The forests release huge quantities of terpenes (unsaturated hydrocarbons) – about 1000 megatons annually, but we do not know why. The tropical forests also used once to cover about 10 per cent of the land surface and their presence affects climate through their ability to evapotranspire water vapour and so sustain the rainy, cooler and cloudy climate of their region.

The desert ecosystem covers much of the remainder of the land. Although life is sparse here the whole ecosystem may, like the ocean deserts of tropical regions, play a significant role in global climate and chemistry. Human-contrived ecosystems cover almost a tenth of the land area, and are increasing. Being transient, they have no established role in Gaia.

The lowly bacteria that sustain all the natural ecosystems fall into three principal groups: photosynthesizers (the analogues of plants), consumers (the analogues of animals), and fermenters. The photosynthesizers are always at the surface of the land or sea, using the energy of sunlight directly to fix carbon dioxide and water to make sugar and release oxygen to the air.

$$\text{Energy} + 6CO_2 + 6H_2O \rightarrow C_6H_{12}O_6 + 6O_2$$

(Sunlight + Carbon dioxide + Water → Sugar/Organic matter + Oxygen)

By removing carbon dioxide from the atmosphere, they tend to cool the planet by removing some of its greenhouse effect (see Chapter 7).

The consumers gain energy by reacting oxygen and organic matter (sugar) made by the photosynthesizers.

$$C_6H_{12}O_6 + 6O_2 \rightarrow 6CO_2 + 6H_2O + \text{Energy}$$

(Sugar/Organic matter + Oxygen → Carbon dioxide + Water + Body energy)

Some of this organic matter, however, escapes oxidation and sinks into the sediments or the soil where the fermenters digest it and convert it to carbon dioxide and methane.

$$C_6H_{12}O_6 \rightarrow 3CO_2 + 3CH_4 + \text{Energy}$$

(Sugar/Organic matter → Carbon dioxide + Methane + Body energy)

A small residue of indigestible organic matter always escapes and is buried in the crust. This burial of a small amount of carbon made by the photosynthesizers is what sustains the oxygen of the air (see p. 111). The part these processes play in regulating atmospheric gases and climate is the subject of Chapters 6 and 7.

These three bacterial groups form the most powerful part of Gaia's body, and have sustained her for some 3.8 billion years. If we in our follies destroy so much of the larger life on Earth, such as the trees and other plants, that homeostasis is threatened, it will be the bacteria that carry on and take over the management of the planet as they have done before.

The ensemble

Biota, biosphere, life, and Gaia. These are words whose meaning is imprecise even to scientists; yet they use them with authority. Quotations from sacred books are held in a similar way to be true by the fundamentally religious, but the fundamentalists can argue that for them truth and authority are a matter of faith. Scientists have no such defence for their ignorance.

The word "biota" is fairly easy to define and most would agree that the biota is the complete catalogue of all living organisms on Earth. But there is still confusion over quality and quantity: is the biota all of the life on Earth, or is it merely the list of all different species of organisms? The word "biosphere" was a geographer's term, originally used by Eduard Suess in the last century to define the region of the Earth where life could be found. The Russian scientist Vladimir Vernadsky used it extensively in his essays on biochemistry to mean something more than mere geographic regions but never as much as the tightly coupled system of Gaia. I have so far never encountered a clear definition of "biosphere". It is a word like "life" that everyone knows the meaning of, instinctively and personally, but cannot define. Even the geographer's use of it to define the region where life is found runs into uncertainty over the boundaries of the undefined term "life".

The definition of Gaia is simpler. Gaia is the planetary life system that includes everything influenced by and influencing the biota. The Gaia system shares with all living organisms the capacity for homeostasis – the regulation of the physical and chemical environment at a level that is favourable for life. We look in more detail at the role of living organisms in maintaining planetary homeostasis in Chapter 7.

Having looked at the bits and pieces of Gaia we need to keep in mind that our planet is forever evolving. It would be misleading to describe the anatomy of a pupated insect, hanging sessile by a silken cord from a twig, without considering its past as a caterpillar or its future as a moth or butterfly. So it is with the Earth – a system that has evolved from a ball of rock to a planet with entirely bacterial life, to the complex ecosystems we now see, and facing an unknown future. The anatomy of the Earth is forever changing.

CHAPTER THREE

Physiology

When I talk of Gaia as a superorganism I do not for a moment have in mind a goddess or some sentient being. I am expressing my intuition that the Earth behaves as a self-regulating system, and that the proper science for its study is physiology. This chapter explores further how physiology can be applied to understanding the Earth and, through a series of computer models of the hypothetical planet Daisyworld, postulates that Gaia, the system of life and its environment, is indeed a superorganism, and capable of self-regulation.

So what is a self-regulating system? The nearest one is you yourself – your body temperature regulation, for instance. But before considering anything as complex as a human (see p. 63), you need to try to understand a simple self-regulating system. Such systems are to be found in nearly every home. In your kitchen, for example, there are likely to be one or more pieces of equipment designed to operate at a constant temperature, such as an oven, an electric iron, or a house-heating system. They always include a source of heat, and a device called a thermostat that senses the temperature. The thermostat can be set to switch off the supply of heat if a chosen temperature is exceeded, and switch it on again if the temperature falls below the chosen value.

An electric iron needs to be hot enough to smooth the damp linen but not so hot that it scorches – say a temperature in the region of 130°C. In days gone by it was good enough to heat the iron on an open fire and use it when it made the right sounding hiss if placed on a damp cloth. It worked, but the careless ironers often scorched their linen because the iron was too hot, or failed to smooth because it was too cold. Something that made fewer demands on the skill of judging temperature was needed, and eventually the electric iron, with a heating element made of high-resistance wire and controlled by a simple "thermostat", was made.

Today's irons work on the same principle. The thermostat acts like a kind of thermometer: a metallic strip that is straight when cold, bends into a curve when heated, the sharpness of the curve increasing as the temperature rises. As the iron heats and the strip bends, one end of it moves away and pulls open an electrical contact. With the contact open no current flows and the iron begins to cool. When the temperature falls sufficiently the strip straightens and the contact is made again, so heating resumes. In practice the iron continues to cool for a short time after the contact closes. This is because it takes time for the heat to flow from the heating element. Similarly, when the contact opens, the iron continues to heat for a brief period before it cools because the heat from the heating element takes a little while to distribute.

Simple though this device is, it illustrates the basic principles of control theory, the circular logic of regulated systems, including those, such as human beings, that are alive. A control system automatically regulates a condition such as temperature through a series of connected functions. It has a sensor, which registers any variation from the set value. It then amplifies and feeds back this error, whether plus or minus (above or below the set point), and initiates a control (turning a switch off or on, for example), which either reverses the variation (negative feedback), or adds to it (positive feedback) (see p. 61). In the electric iron, for instance, sensor, amplifier and controller are all one entity – the bimetal strip. The iron regulates by negative feedback, and its response is quite slow. Negative feedback can be quite sluggish in restoring a desired state. Positive feedback is faster, but unstable – it can lead to a "runaway" state. Natural control systems often combine both, for successful self-regulation.

Living organisms, and control systems generally, exist only where there is plenty of free energy. This may be energy from sunlight, chemical potential energy (stored in plants), or a supply of heat and power. Such systems are also affected by their environment. The iron, for example, cannot regulate if the environment is hotter than the chosen temperature; and if the power supply is limited, it will be unable to reach the desired temperature in cold environments. But the intriguing feature of the iron is that, in cold environments, it regulates on the hot side of the chosen level, and in warm environments, it regulates on the cool side.

A device such as an electric iron is simple enough to understand in human terms. But if you try to explain how it works in the scholastic way of cause-and-effect logic, you will find the explanation exceedingly difficult, if not impossible, to formulate or understand. You might think that the mathematical analysis, the reduction, of such a simple device as an iron, would require no more than high-school mathematical skill. In fact the analysis by traditional calculus of these self-regulating systems is always impossibly complex and even an approximate solution is beyond most students. Classical mathematics is essentially reductionist, and so is ill-equipped to deal with complex natural systems. There are many things in our daily experience, from riding a bicycle to catching a ball, that defy explanation without

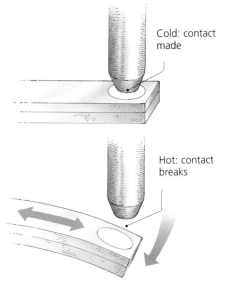

Cold: contact made

Hot: contact breaks

Bimetal strip

Thermostats often use a bimetal strip. Two thin strips, each from a metal with different expansion properties on heating, are welded together. When cold this bimetal strip is in contact with the switch, which turns on the heat. But as it heats, the strip will bend in the direction away from the side with the more expansive metal, to the position where there is least strain at any given temperature. As it bends away, contact is broken and the heat input ceases.

Temperature control in an electric iron

The sensor of an electric iron is a bimetal strip. From information on the actual and the set temperature, it provides the answer to the question: is the iron too hot? The strip is also the controller, and a powerful amplifier, since at a small difference in temperature it can turn the full heating power on or off. The graph, below, shows the cycle of heating and cooling around the chosen or set temperature for the iron. The sawtoothed swing of temperature is characteristic of regulation by a simple thermostat of this kind. No matter how well made a control system, there is always an error, a variation, around the chosen value. Living organisms and systems are never perfect. Indeed error is an essential part of real systems. It is the difference between the perceived temperature and the set temperature – that is, the error – that is amplified by the system to provide a corrective feedback. If the system were perfect, and there never was an error of this kind, the feedback and control would be unnecessary.

How the temperature of the iron is affected by ambient temperature

The curve below shows how in cold environments, the iron would always be slightly hotter than the set point, and in warm environments it would tend to be slightly cooler. This is because the amplified feedback that aims to correct the iron's temperature "overshoots" making the iron too hot if it has sensed it to be very cool, and vice versa. Daisyworld shows this kind of over-correction – as does a drunken driver trying to negotiate a bend!

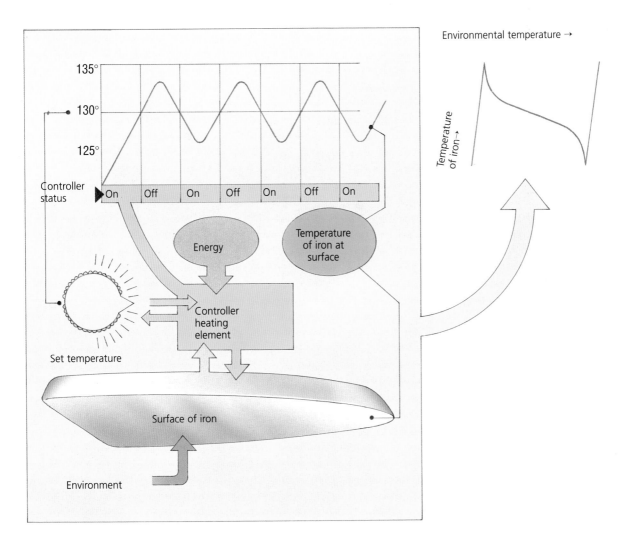

access to a powerful computer, and some not even then. The famous physicist James Clerk Maxwell (1831–79), who first reduced to elegant mathematics the theory of electricity and magnetism, is said to have been baffled by the complexity of the steam engine governor, an invention of his fellow Scot, James Watt (1736–1819).

I well recall, as a child, seeing one in operation at the Science Museum in London, England. It seemed so simple, so obvious. Yet so narrow are the constraints of reductionist analysis that even Maxwell could not immediately solve the expression of its operation in classical mathematical terms.

If simple mechanical control systems are so difficult to interpret, what chance have we of explaining how living organisms work? The most impressive text I have so far read on the control theory as applied to organisms, is by D S Riggs: *Control Theory and Physiological Feedback Mechanisms*. First published in 1970, long before computers were freely available, it remains an elegant and evergreen exposition of the great problem of expressing in simple yet succinct terms the first steps in understanding the behaviour of living systems.

In the introduction to this book, Riggs explains why control theory is rarely used by biologists. He stresses two main difficulties. First, most students of biology have in the past tended to be recruited from those who were scientifically inclined but found mathematics difficult. Second, the most developed and easily comprehended methods of control theory deal with what engineers call "linear systems". A linear system is one where the interactions between the component parts can always be expressed in terms of simple arithmetic – for example, calculating the time for filling and emptying a tank of water, where the flow is controlled by a simple float that opens and closes a valve. Compare that with the complex interrelated nonlinear systems that regulate the quantity of water held in your body. Unfortunately nature is inherently nonlinear, and never more so than in living systems. This hindrance is intrinsic, and it is no use trying to sneak around it by applying inappropriate linear methods to grossly nonlinear systems, simply because linear methods are easy to use.

Not only scientists, but all of us, are so constrained in every walk of life to think in a linear logical way, in terms of cause and effect, that we find it difficult to understand the circular logic needed to explain the workings of living organisms or even of a non-living system such as an automatic pilot. We have been dazzled for generations by the mathematical physicist, the genius who can, with fluent flourishes, cover a blackboard with equations that explain the workings of all levels of the universe from the ultramicroscopic subatomic particles to the vast reaches of space. Such eloquence can convince us that the physicist is a friend of, or at least closely acquainted with, God. It can be as entertaining as a good concert pianist and its hubris is charmingly disguised as harmless modesty.

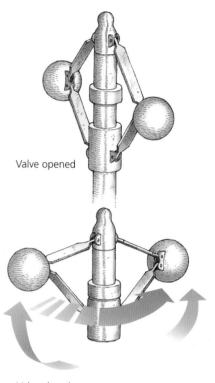

Valve opened

Valve closed

Steam engine governor

The steam engine governor is one of the best known simple control systems. A steam escape valve is housed in a rotating shaft, driven by the engine. Two heavy metal balls hang on hinged arms attached both to the top of the valve and the shaft (see above). Before the engine starts the shaft is still and the balls hang down; the valve is open (top). As it gets up steam the engine increases speed, the shaft rotates faster, the balls swing up, and the valve is closed (bottom). Steam is shut off, and the engine slows again. So the speed of the engine is regulated.

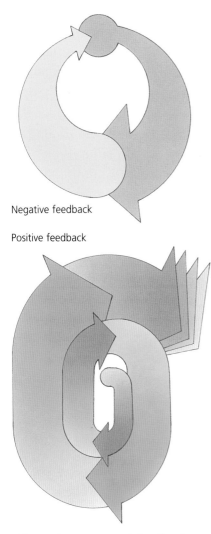

Negative feedback

Positive feedback

Control systems and feedback

A control system has three linked functions. It acts as a *sensor*, detecting the condition being regulated (such as temperature) and registering any variation from a set level. It is also an *amplifier* that magnifies the initial variation, and a *controller* of the inputs to the system. The amplified variation triggers a feedback control: either *negative feedback* (top) reversing/ inverting the change, or *positive feedback*, amplifying it further (bottom). Amplification is essential in any self-regulating system – without it neither positive nor negative feedback can have any effect.

The sheer brilliance of these explanations of the nature of matter and the universe has, sadly, led us to believe that their reductionist view of the universe is sufficient in itself.

An even greater hindrance is our mental equipment, something that has marvellously evolved to allow us to solve the most intricate nonlinear mathematical problems without us ever understanding how. Consider, for example, a game of tennis, where the two players have to react instantly to the ball as it hurtles between them. Quite unconsciously their minds accept the brief moving signal from their eyes and compute the course of the ball, while simultaneously directing the movement of their arms and rackets so as to intercept the ball at the very point in time and space that the racket and ball coincide. To explain that simple act in mathematical terms requires a vast assemblage of nonlinear equations. The tennis players, in effect, solve them almost instantly in the act of hitting the ball.

Biologists, like tennis players, know a lot about the game they play with living organisms, but like the tennis players, they can neither understand nor explain what it is they do so well. This is why biologists cannot or will not try to define life.

Circular logic exists and is used, but only in the difficult sciences of physiology, control theory in engineering, and in computer science. A few years ago the need for a holistic science that would include all of these led Norbert Weiner to coin the term "cybernetics" for a general science of self-regulating systems. It has not been a success. Cybernetics still tends to be a Cinderella among the sciences – along with engineering itself – subjects fit for "rude mechanicals" able to develop exciting inventions, but unable to explain in classic linear logic how their inventions work.

In the 30 years since Riggs' book was published, and in spite of the ubiquity of computers, little further progress into the understanding of living organisms as systems has been made. But physiologists are like engineers and are never afraid of using a top-down approach to systems; as long ago as the last century the great French physiologist, Claude Bernard (1813–78), was well aware of the circularity of the notion of feedback in the self-regulation of the body. Later the American physiologist Walter Cannon (1871–1945), coined the term "homeostasis" to express the wisdom by which living things kept their constant comfortable state when faced with internal or external change. Since I wrote the first edition of this book in 1990 the mathematician Peter Saunders of King's College, London University, has analysed the Daisyworld model and thrown such new light on the mathematics of self-regulation that it inspired him and a South African physiologist, Johan Koeslag, to apply the Daisyworld model to the real human physiological problem of blood glucose regulation, with stunning success.

The open-minded acceptance by physiologists that the linear logic of the rest of science was unhelpful in their quest to understand the body allowed them to make the first steps toward explaining how living organisms work. A fine example of their way of science is illustrated by the physiology of temperature regulation – how we as humans keep the core regions of our bodies at constant temperature over a wide range of environmental temperatures. Until comparatively recently it was believed that there was within the body a temperature sensor with a reference point set near 38°C. Any rise or fall in core temperature from this reference point would be recognized and the appropriate action taken: increased metabolism and shivering if cold; sweating and an enlargement of the skin blood vessels if hot. T H Benzinger and his colleagues, in an elegant investigation, have shown that this simplistic view of body temperature regulation is wrong. There is no formal reference point; instead there is a kind of automatic consensus among the systems of organs that constitute the body as to what is the best temperature. Temperature regulation takes place as a result of the tight coupling of five processes: metabolism; sweating; blood vessel dilation; core-modulated shivering, and skin-modulated shivering.

Control theory and planet Daisyworld

The physiologists are part way toward understanding the application of control theory to living organisms. Is it possible to use this same approach with the Earth, assuming that it also is a form of living system? One way to do this came unexpectedly from a model that started out as no more than an answer to criticisms of Gaia theory by the biologists, W Ford Doolittle and Richard Dawkins. Both have criticized the theory on the grounds that there is no way that the diverse living organisms of the Earth could act altruistically (to their mutual advantage) to regulate the planetary environment. Evolution by natural selection, they say, would always be in favour of the genes. Symbiosis, they note, is rare and only between closely related organisms. Both were scathing in their criticisms. Doolittle observed that planetary regulation would require foresight and planning by the biota, the meeting of committees of species to negotiate next year's temperature. Dawkins claimed that Gaia could not exist because there was no possibility of the Earth reproducing and therefore no possibility of evolution by natural selection among the planets!

These criticisms were valuable: they made me realize that I had been thinking intuitively rather than rationally about Gaia and that a proper explanation was lacking. One way to answer them was to express what I meant by Gaia in the form of a scientific model. I called my model "Daisyworld" because it was quite simply a computer model planet on which grew just two species of organisms, dark- and light-coloured daisies.

Daisyworld can be imagined as a planet like the Earth orbiting a star just like the Sun. The surface temperature of a planet receiving sunlight is affected by its depth of colour because of the albedo effect

Human temperature regulation

Human body temperature is kept at an optimum, not by some hidden sensor programmed to a set point, but by a kind of automatic consensus arrived at by the brain and the various organs and systems of the body. When the ambient temperature becomes too hot or too cold, our bodies respond by either sweating or shivering, burning food or fat, and reducing or increasing the rate of blood flow to the skin. The graph, right, is an engineer's diagram showing the power of the five processes of human temperature self-regulation to function when a naked human is exposed to a range of different ambient temperatures.

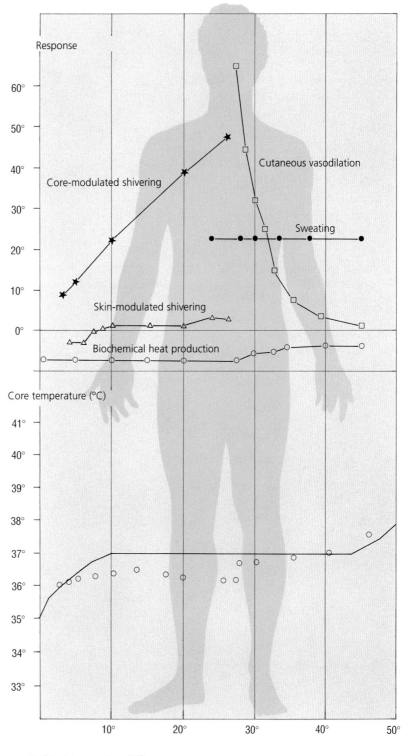

Observed and calculated core temperatures compared

It is the core temperature of the human body that is closely regulated, remaining around the "normal" temperature of 37ºC. (Skin, hands and feet can tolerate a far greater temperature fluctuation.) If you compare the core temperature of a real human (solid line on the graph, right) with the temperature you would expect from the information given in the engineer's diagram above (summarized by the dots, right), you will see that the match is close. Thus it is possible to account for human temperature regulation by a consensus result of the five different response systems.

What is a system?

A system may be defined as an assemblage of objects (or ideas) linked together by some form of regular action or interdependence. This is what we have in mind when we think of natural systems such as the Solar System, the nervous system of the human body or the Earth – or of human systems such as a transport system, or an economic system. The component parts of any of these real systems are numerous and the interactions between their component parts very complex. Biological systems are far more complex than physical ones such as galaxies, or particles, and waves. Scientists can never hope to analyse complex systems in full. Instead they are obliged to begin by introducing simplifying assumptions, which enable them to represent the real system by an abstract model.

Waves and particles
Sub-atomic systems

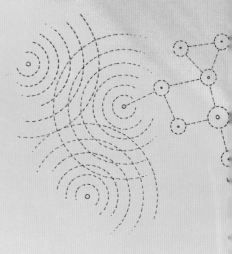

Fragments for life
Molecular systems

What is a model?

"A model, like a map, cannot show everything. If it did, it would not be a model but a duplicate. Thus, the classic definition of art as 'the purgation of superfluities' also applies to models. The model maker's problem is to distinguish the superfluous from the essential."

D S Riggs

A model is a representation of a real system that is simple enough to describe in detail but still retains some of the essential features of the real system. A good model must behave sufficiently like the real system to allow fairly accurate predictions about that real system's behaviour. A familiar example of a simple model is the "ideal" gas of the physicists, made up of perfect atoms that behave like microscopic billiard balls. Physiology is concerned with much more active and complex systems; its models will nearly always now be simulations built in the information space of a computer.

Termites to Gaia: superorganisms

A superorganism is an ensemble of living and non-living matter, which acts as a single self-regulating system. The organisms comprising the living part, and the material, non-living part, are tightly coupled in at least one environmental variable. A beehive, for instance, is made of insects and a wooden enclosure; this ensemble is able to regulate its internal temperature thermostatically almost as well as a mammal. A termite colony is another example. Ecosystems are superorganisms – from the anaerobic microbial communities to the great tropical forests (see Chapter 2). Gaia is the largest and most complex superorganism we know.

Dust in the universe
Solar, stellar, and galactic
systems

Viruses to redwoods
Organisms

Termites to Gaia
Superorganisms

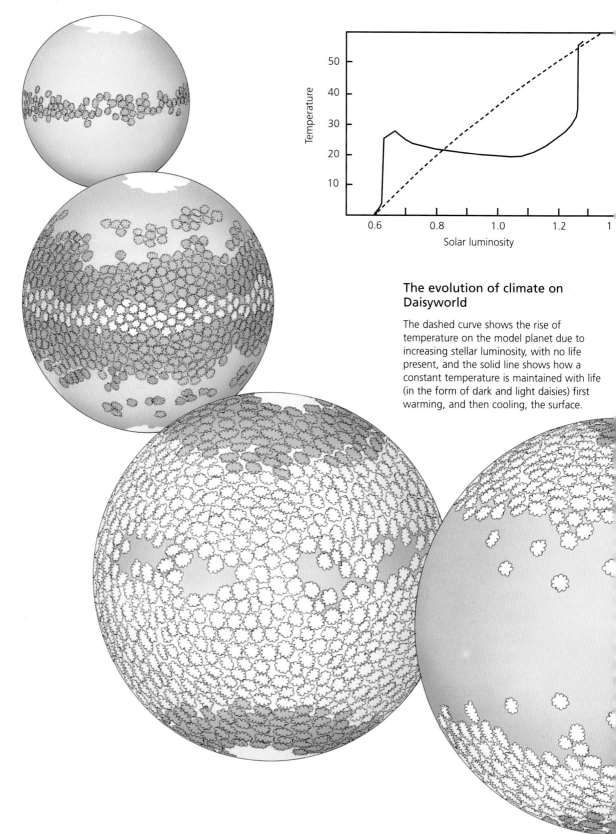

The evolution of climate on Daisyworld

The dashed curve shows the rise of temperature on the model planet due to increasing stellar luminosity, with no life present, and the solid line shows how a constant temperature is maintained with life (in the form of dark and light daisies) first warming, and then cooling, the surface.

Daisyworld: a model for Gaia

Welcome to planet Daisyworld: a computer model planet like the Earth, orbiting a star like the Sun, but on which the only species are light and dark daisies. In the distant past, when the star was less luminous, only the equatorial region would have been warm enough to permit the growth of daisies, and the dark daisies would have flourished, because they absorb more warmth from sunlight. Gradually the dark daisies would have colonized most of the planet, and by absorbing heat begun to warm the surface environment.

However, as the star's luminosity increased the lighter daisies would have been favoured instead, due to their natural ability to keep themselves and the planet cool, by reflecting more light. Finally, when the heat flux from the star becomes so great that not even the white daisies can keep the planet cool enough for life, deserts spread from the equator and finally the system fails and Daisyworld dies.

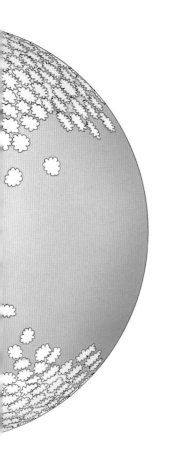

(see p. 147): a darker planet absorbs sunlight, and so warms up. Daisyworld's bare ground is mid-toned, neither dark nor light. Our model planet is moist and fertile; seeds will grow whenever the temperature is warm enough, above 5°C. Growth is best at temperatures near 22°C, but declines at higher temperatures and ceases above 40°C. Imagine that, just like the Sun, Daisyworld's star warms up as it evolves. Would the evolution of the Daisyworld ecosystem lead to the self-regulation of climate?

When the star has grown in output sufficiently to raise some of the equatorial regions of Daisyworld to 5°C, a few daisy seeds germinate and grow. Dark daisies are favoured because they absorb more of the star's light and are warmer than the bare earth. Light daisies are discouraged because they reflect starlight and are cooler than the bare surface. At the end of the first season there are many more dark daisy seeds left on the ground than light daisy seeds. At the start of the next season, dark daisies are blooming in profusion and light daisies are only rarely seen. The extensive ground cover by dark daisies alters the surface reflectivity, and the planet is warmed as well as the daisies. Soon, with a powerful positive feedback, dark daisy growth and planetary temperature rise until at some point above the optimum for daisy growth, temperature and growth level off. Now that the planet is warm, light-coloured daisies grow and compete for space with the dark daisies, until their cooling effect adversely alters the environmental temperature. Ultimately a steady state of dark- and light-coloured daisies is reached, with a mean albedo close to that needed to maintain the surface temperature at a value optimal for daisy growth.

As the star increases its heat output so the dark daisies decline and the population of light-coloured daisies spreads. The temperature remains close to that preferred by daisies. Like the electric iron, the system tends to regulate on the hot side of optimum when the starlight is weak at the beginning, and on the cool side when the output of the star is great, later on. Eventually in the evolution of the star, its output grows so great that even a planet-wide cover of the heat-reflecting light-coloured daisies is insufficient to keep a tolerable climate for daisies; the system suddenly and catastrophically fails, and Daisyworld dies.

The simple model Daisyworld is the merest caricature of Gaia. It concerns the regulation of a single variable, temperature, and includes a single species, daisies. Even so, it shows a likeness to the natural systems of living organisms, like temperature regulation in humans. It regulates strongly, is robust, and is resistant to perturbation. Like the human temperature regulation system, it has no formal single set point of reference as a goal for regulation; in this respect it differs from engineering systems whose thermostats refer always to such a set point. Daisyworld works by evolving a climate that suits the preference of its daisies.

The model Daisyworld was made to answer the biologists' criticism that global temperature regulation, involving living organisms, would invoke foresight and planning by the organisms, which is absurd as well as impossible. Daisyworld works wonderfully well and refutes the criticism; Gaia can regulate without the need for foresight or planning by the biota. The regulation is entirely automatic.

But the Daisyworld model has turned out to be very much more than the answer to an inadequate criticism. In spite of its simplicity, Daisyworld is a complete example of a new kind of global model, a geophysiological model. A model in which the evolution of the organisms and the evolution of their physical environment are tightly coupled together so as to constitute a single evolutionary process. Such a model could not easily have arisen from the fragmented body of knowledge that is science now. Biologists, unfamiliar with geophysics and geochemistry, do not make models that include the material environment. Geophysicists and geochemists do not include organisms in their models. Even biogeochemists, who recognize the existence of the biota, fail to do more than include organisms in boxes, loosely coupled, and unresponsive.

I am often asked what is new in Gaia theory that was not already expressed by the father of biogeochemistry, the Russian scientist, Vladimir Vernadsky, and by the discipline's greatest exponent, G E Hutchinson. The simple answer is that biogeochemistry differs from geophysiology in the same way that a platonic friendship differs from a happy marriage. In biogeochemistry, organisms and their material environment are recognized as coexisting and coevolving but still separately, like friends. In geophysiology, the organisms and their environment are so tightly coupled that they constitute a single system, Gaia. The evolutionary system Gaia is an emergent domain, something of which the whole is more than the sum of the parts. In this it is like the domain of a good marriage. That mainstream science still thinks in separated terms is well illustrated by the acceptance of the title International Geosphere Biosphere Program (IGBP) for the principal global scientific programme. As a geophysiologist I cannot accept that the geosphere and the biosphere can properly be investigated as if they are separate and independent. As I revise this book in the 21st century, I am pleased to record that two scientists, who were recently prominent in the running of the IGBP, Chris Rapley and Peter Liss, both tell me that the inappropriate name was an accident of history and that the IGBP now represents first-class Earth system science.

Attempts to sabotage Daisyworld

There have been a number of attempts to dismiss Gaia by seeking conditions under which the Daisyworld model does not work. I welcome these since it shows that their instigators recognize that the model is in many ways the key to Gaia and to understanding

Planetesimal impact

The impact from a planetesimal can devastating. Obviously, the bigger th meteorite, the greater the damage. has been reckoned that even a comparatively small planetesimal, so 10 kilometres in diameter, can cause crater 300 kilometres across, splashi molten rock and gas far out into spa

CASE HISTORY Marasmus

In human medicine a condition of malignant debilitation sometimes occurs after severe shock. In this condition, called marasmus, metabolism and normal physiological function fade away until the power to conduct even the minimum homeostasis needed for life is lost. It is a condition that used to afflict some women who had suffered a particularly difficult childbirth accompanied by a significant loss of blood.

Gaia as a system, on at least one occasion, has come close to a fatal marasmus in the aftermath of a disease or injury, perhaps a severe planetesimal impact. About 65 million years ago there was an extinction event where 90 per cent or more of species present in the rock record suddenly disappeared (see also pp. 145–6).

The success of Gaia as a system requires the continued and active maintenance of the environment within the range suitable for life. This requires more than some minimum total biota, for there must be enough living organisms to offset by their existence the remorseless evolution of the planet toward chemical equilibrium. I sometimes say that there can be no sparse life on a planet. The system must be large and vigorous and able to handle planet-sized problems. Invalid planets soon die.

If life were reduced to one-thousandth of its present abundance and kept there, the Earth would in a few million years evolve into an exceedingly inhospitable place. Carbon dioxide would rise in abundance to several per cent. Oxygen and some nitrogen would leave the atmosphere, and the oceans would become more salt – perhaps too salt for life to restart. The temperature might become very high, perhaps in excess of 30°C as a mean. This would partly be a consequence of increased carbon dioxide, but it could also come from a change in cloud density and structure. The present clouds above the oceans may owe their existence to the marine algae. Some of these emit sulphur gases that are the source of the nuclei without which clouds cannot form (see p. 127).

The longer life stayed sparse, the more difficult would be the restoration of the Gaian system. The Sun's heat is in the longer run always increasing, as is the abundance of carbon dioxide and the salinity of the oceans. Perhaps the most serious change would be the loss of water, as hydrogen began to escape from an Earth that had returned to the oxygen-free state (see pp. 80–1).

It is a tribute to the powers of recovery of Gaia that none of the 30 or more planetesimal impacts has ever caused a fatal marasmus.

geophysiology. Before considering these attempts to overthrow Daisyworld, it is necessary to recall that the original purpose of Daisyworld was to show that climate regulation on a global scale could occur without invoking foresight or planning and also, in the simple constrained ecosystem of dark and light daisies, without violating the evolution of the daisies by natural selection.

The first attempt to demolish Daisyworld was to propose that in a real world there would be neutral-coloured daisies whose growth had no effect on climate. Furthermore, since these daisies would be relieved of the need to synthesize pigment, they would have an advantage over the coloured daisies and would displace them. These cheating daisies who would not play the game would stop the temperature regulation, which depended on two species of daisies competing for space. The model would revert to the conventional wisdom of a climate determined by geophysics only.

It is simple to add to the Daisyworld model an additional species, neutral-coloured daisies, that have a 5 per cent faster growth rate because they do not need to synthesize pigment. The graphs, right, show that the neutral daisies do not impede temperature regulation, nor do they supplant the dark- and light-coloured daisies. (The temperature/solar luminosity graph shows the temperature regulation by the daisies (see p. 66).) What happens is that when the heat received from the Daisyworld star is weak, only dark daisies are warm enough to grow and leave seeds. When the star is hot, only light-coloured daisies grow. The only condition that permits the growth of the neutral daisies is the absence of a need for regulation. When the star's output is just right for life on Daisyworld, only then is the existence of the cheats affordable.

There have been several other attempts to deny the validity of Daisyworld as a model of a geophysiological system. Some critics used the fact that there is a limit to the range of solar luminosities over which this self-regulating system can exist as evidence that the idea of self-regulation is flawed. To these critics, I can only say that the fact that you may die if you are overheated or frozen does not mean that you are not alive now. Other critics made Daisyworlds that included viruses that destroyed the daisies – they saw the death of the system as an indication that in a real world regulation could not occur. Again I reply that the death of a system is surely proof only that it is mortal, not that it does not exist.

Daisyworld was deliberately restricted in the first model to two species to make its point. But in a real world, many species of organisms and mutations would be possible. The graphs, far right, show an evolutionary model where the daisies can adapt by changing their colour to that best fitted to the heat flux from the star. In addition, there are herbivores (rabbits that eat the daisies) and predators (foxes that cull the rabbits). The model is also perturbed on four occasions by events that suddenly destroy 30 per cent of the daisies.

Effects of cheats on Daisyworld

Critics of Daisyworld argued that if unpigmented or grey daisies (which do not have to expend energy synthesizing pigment and so grow faster) were added, the model would not work.

The second model incorporated cheating grey daisies. A 5 per cent

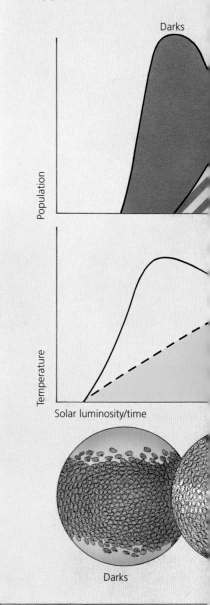

Darks

Population

Temperature

Solar luminosity/time

Darks

growth tax was imposed on the coloured daisies for making pigment. The graphs show that grey daisies flourish only when conditions are optimal. They do not prevent temperature regulation by the light and dark daisies.

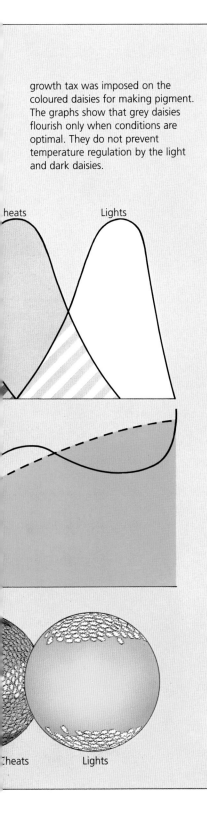

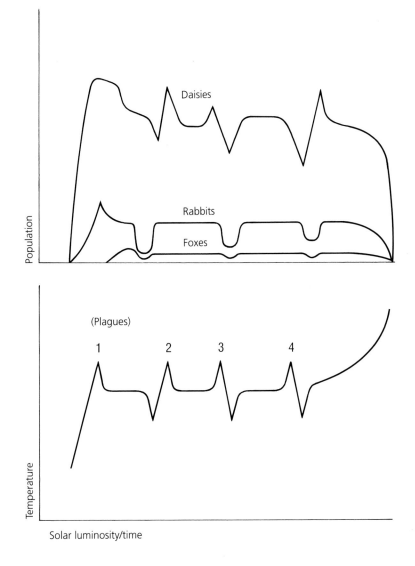

Foxes, rabbits, and daisies

In an attempt to simulate a real world, herbivores (rabbits) and predators (foxes) are introduced into the Daisyworld model. The model is also subjected to periodic catastrophes: on four occasions 30 per cent of the daisy population is destroyed by a plague (see above). Surprisingly, neither the addition of herbivores and predators, nor plagues, seriously affect the daisies' ability to regulate climate. The crashes of daisy populations, and the corresponding dips in herbivores and predators, all are short lived. So too are the dips in temperature regulation, from which the system rapidly recovers. In fact, this model exhibits a remarkable robustness to perturbation, which only fails when solar output becomes too great. Geophysical models such as this are stable because the populations are constrained as they would be in the real world, by environmental conditions.

The adaptive evolutionary Daisyworld is just as good at temperature regulation as the original model where there are two species of daisies with fixed colours. The existence of herbivores and predators does not seem to impede the model's capacity to regulate. Indeed, the regulation is robust and resists substantial perturbations.

An intriguing feature of the model with the cheating grey daisies and the model with the herbivores and predators is that, contrary to the usual experience of population biology models, more than two species of organism are present and coexist in a stable steady state. Prior to the Daisyworld model, population biology was almost always limited to two species for its models. As in the famous three-body problem of astrophysics, the addition of a third species, like the addition of a third body, rendered the system unstable and tending to chaos. The tight coupling of Daisyworld allows both positive and negative feedback to occur constructively and as stabilizing rather than perturbing influences.

So far I have been considering models, which are only abstractions of the real world. The Daisyworld model successfully demonstrates that physiological self-regulation may be an emergent property of the system Gaia. But you may well ask, what evidence is there that the real world behaves as a physiological system, and does thinking about it in that way help with understanding the past and the present and predicting the future? In Chapters 4, 5, 6, and 7, I present the evidence gathered so far. You must draw your own conclusions.

During the ten years of the 1990s Daisyworld models have evolved in scope and complexity thanks to my colleagues, Tim Lenton and Stephan Harding, and to the acceptance of Daisyworld as a template for large climate and Earth system models by scientists at the Hadley Centre in the UK and at the Potsdam Institute in Germany.

CHAPTER FOUR

Epigenesis

"Fair daffodils we weep to see you haste away so soon
As yet the early rising sun has not attained his noon."

In these famous lines, Robert Herrick takes notice of the difference in time span of the lives of daffodils and poets. Poets and the rest of us are long-lived compared with flowers. But compared with Gaia, who has lived for 3.8 billion years, our lifespan is brief.

This chapter looks at how Gaia came into existence, and the turbulent period of her early life history – Gaia's childhood, if you like. But before we can properly discuss these momentous events, it is worth pausing to consider the time dimension, for the birth of Gaia happened so long ago that the timescale is almost beyond our comprehension. In fact, the time dimension is probably the most significant way in which Gaia differs from other forms of life. Assuming that time began with the Big Bang more than 10 billion years ago (see p. 77), Gaia has lived 3.8 billion years – about one third of the age of time itself. Yet the life of a bacterium, one of the simplest life forms, can be measured in hours or minutes!

My immediate problem, therefore, in comparing Gaia with other living organisms, is to express times ranging from the age of the universe to the shortest interval that humans can perceive, about one tenth of a second, and to express them on the same diagram. For the illustration overleaf, I am grateful to Nigel Calder, author of the splendid book *Timescale*. In this book, he devised an elegant and economic way of expressing time by dividing it into intervals, each one ten times longer than the one before. These intervals were then displayed as a series of segments wrapped in a spiral moving outward from the beginning at the centre. In this way he was able to express in a single diagram the course of time from its origin, through the Big Bang and the amazing events of the first trillionths of a second, on to the formation of matter, of elements, and of galaxies, right up until now, more than 10 billion years later. But since our interest is in the living organisms, and not in the inanimate universe, my scale will be reversed and start from now and go back in time. I do it this way because the short intervals are in the present, like those of our own perceptions, and the long ones are the ages of Gaia and of life on the Earth.

As I have already mentioned, bacteria have the shortest life spans. At first it is not easy to see how the lifespan of a bacterium is determined at all. Bacteria multiply by splitting in two, and both new bacteria after a decent interval, say 20 minutes, split again and so on. One way to measure the lifespan of a bacterium is to let bacteria grow with a limited food supply until eventually a steady population is reached, with bacteria dying at the same rate as new ones appear by division. It is possible, by counting the total number of bacteria in a small volume at intervals as they grow toward the steady state and comparing this number with the number of viable organisms, to determine the lifespan of a bacterium. Like all other living things, it varies according to the quality of life, but under favourable circumstance the lifespan of a bacterium can be as much as a few days.

Larger individual living organisms have lifespans ranging from days for some insects to thousands of years for some large trees. There is a rough relationship between lifespan and the rate of consumption of oxygen per gram of tissue. The larger the throughput of oxygen, the shorter the life. The biologist Dr Thomas takes the view that breathing oxygen is equivalent to exposure to radiation (see p. 114). To me the most astonishing thing shown on this scale is the longevity of life itself. On this logarithmic scale it is in the same sector as the age of the Earth, the Sun, and the galaxy.

Let us now go back and look at the birth of life and of Gaia. But first we have to go back 4.6 billion years, to the origin of the Earth and the Sun. To understand the birth of our living planet, we need to know the cosmic environment in which it evolved.

The birth of Gaia

Some 4.6 billion years ago a supernova exploded somewhere close to the cloud of gas from which the Sun and Earth were formed. Supernovas were said by astrophysicists to be star-sized thermonuclear bombs. Their explosions are so vast that in the ten seconds of their occurrence the output of energy is as great as that from all the rest of our galaxy. Modern astronomers think that it took an even larger explosion to start the universe itself. With their telescopes astronomers can see the stars and galaxies and can measure their motion through space. They see the visible part of the universe expanding as if it all came from a single point some 12 to 15 billion years ago. For all the vast mass of the universe to keep on moving out against the pull of its own gravitational attraction implies that at the beginning there must have been a very big bang indeed.

If it took a big bang to start the universe, how did a supernova come to start life on Earth? First and foremost the explosion perturbed, and contaminated with fallout, a nearby cloud of hydrogen, helium, and a few other gases. The perturbation caused

Lifespans

The various forms of life have an amazing range of lifespans, as suggested by the scale in powers of ten below. Life on Earth, and Gaia, have survived for 3.8 billion years. Yet the members of a colony of bacteria may have individual lifetimes of only hours or days. Flowers and insects have so short a lifespan that our mere 70 years seems almost forever. Small mammals live only a year or so, larger mammals such as cats 15–20 years. Our lives are brief compared with those of most trees. Compared with the age of life itself, we live but a transient moment.

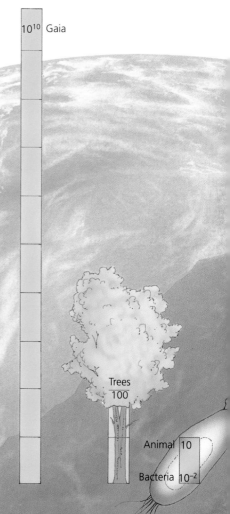

10^{10} Gaia

Trees $\dfrac{}{100}$

Animal 10

Bacteria 10^{-2}

Ages of life

The illustration (right) shows the progression of time, and on it are marked the ages of origin of various life forms. Also for reference the ages of major features of the physical universe are marked. The scale is logarithmic, that is, it is in powers of ten. It spirals back from the present (any convenient time, say 2000) to the start of the universe, each full turn of the spiral down being 10^2 (=100) times longer than the one before. It only takes six such turns of the spiral to go back from a few days ago (10^{-2} years) to the start of the universe, more than 10 billion years ago, when time began. On this scale the great age of life and Gaia can be seen clearly: about one third as old as time itself.

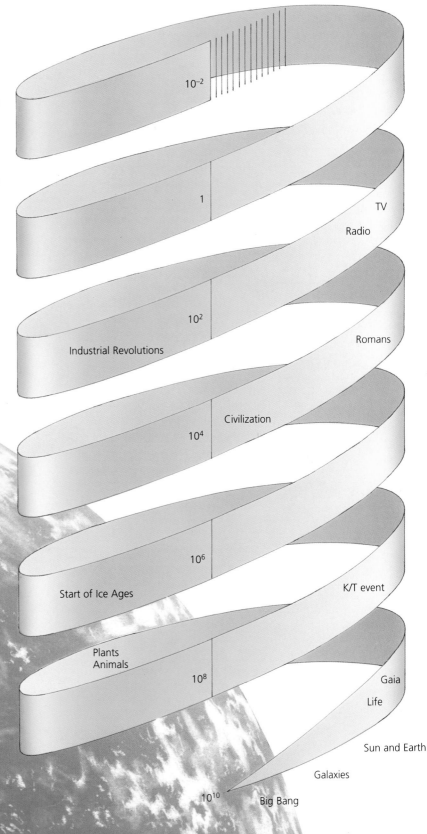

10^{-2}

1

TV

Radio

10^2

Industrial Revolutions

Romans

Civilization

10^4

10^6

Start of Ice Ages

K/T event

Plants
Animals

10^8

Gaia

Life

Sun and Earth

Galaxies

10^{10} Big Bang

some of the cloud to commence condensing under its own weight to form the Sun and the planets. (For a first-rate account of a supernova and the events leading up to it, read the article by Stan Woosley and Tom Weaver in the August 1989 issue of *Scientific American*.)

Compared with a hydrogen bomb, the supernova was amazingly dirty. The fallout was so radioactive that some astronomers think that most of the light and heat of the first few months of the outburst was due to the decay of radioactive nickel and cobalt, made in the explosion itself. It was so radioactive that even 4.5 billion years after, there is in each one of us 30,000 becquerels of radioactive potassium, or about 500 becquerels per kilogram of body weight. If this level of radioactivity of a similar radioactive element, cesium, were present in meat, it would be enough to condemn it as inedible. This is a reason for comfort rather than concern when considering the dangers of radioactivity, for we have all evolved in a radioactive world. Indeed, we are radioactive naturally, and inhabit a nuclear-powered universe in which, once in a while, one of the nice safe power plants, a star, blows with a spectacular bang. Just as well they do, else we would not be here.

The Earth when it aggregated from the cloud of dust and gas around the newly formed Sun was rich in the fallout of the supernova. The gases of the cloud were captured mainly by the superior gravity of the Sun itself and the larger planets such as Jupiter and Saturn. Earth, Mars, and Venus, which are solid, were made from elements synthesized in the star that exploded. Among these elements were many essential for life, such as carbon, oxygen, nitrogen, sulphur, phosphorus, and iron. The inner planets of Mars, Earth and Venus were also swept by fierce winds coming from the Sun as it settled down into its long stable state, and these winds carried away any early atmosphere that the Earth might have possessed.

Thus the Earth began as a planet-sized chunk of radioactive fallout from a very large thermonuclear explosion. The final act that shaped the present Earth may have been a collision with another planet, about as large as Mars and travelling at a speed of at least 16,000 kilometres per hour. The energy released by the collision melted the Earth and detached a chunk of molten rock that is now the Moon. The melting of the Earth ensured the segregation of the rocks so that the heavier ones such as metallic iron and nickel fell to the centre, while the lighter rocks floated on top and eventually cooled to form a crust.

The scene was now set some 4.6 billion years ago for life to appear on Earth. But first the planet had to cool and to gain a new atmosphere, and the Solar System itself needed tidying up. There were many chunks of rock, smaller than planets, but still kilometres in diameter, orbiting the Sun. These were scavenged by collisions with the planets. The surface of the Moon carries a record of some of the early collisions.

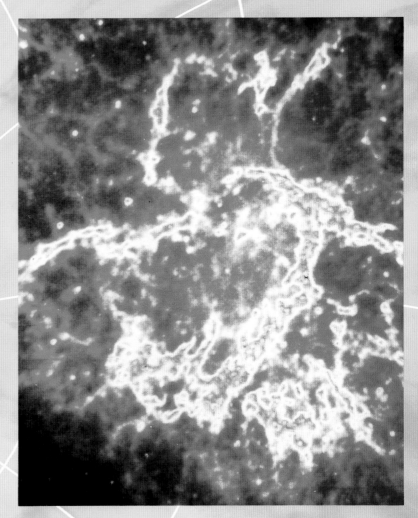

The Big Bang

There is a consensus among astronomers that space and time began more than 10 billion years ago in a near infinitely powerful explosion taking place in a volume smaller than that of the nucleus of an atom. For once they are lost for superlatives and just call it the Big Bang. The whole universe was compressed into that ultramicroscopic bomb and has been exploding outward ever since and cooling as it expands. The hot early moments were no place for life or even for chemistry. Life, with its vulnerable chemistry, had to wait until there were planets cool enough and stars bright with the premium energy needed for photosynthesis.

Big Bang echoes

As the universe expanded and cooled, the hierarchy of particles condensed from the hot primeval plasma – at first quarks and their companions; later protons, neutrons, and electrons; later still elements and finally matter as we know it and the vast variety of chemical compounds. Along with matter there were the particles of radiant energy, the photons (light). At the beginning the energy of the photons would have been measured in giga electron volts. Over 10 billion years later, it has cooled to almost absolute zero – the still-lingering echo of that titanic explosion from which the universe originated.

The Crab Nebula

This cloud of dust 15 light years long and 10 light years across is the remnant of a catastrophic supernova explosion, which was observed by Chinese and Japanese astronomers in 1054AD. The remains of the star that exploded is the Crab pulsar, a neutron star only 20 kilometres across. It was a supernova explosion such as this that provided the energy and matter for life on Earth.

Hadean
4.6–3.7 billion years ago

The period on Earth before life began is well named the Hadean because of its fiery violence. The Earth was far more radioactive than now, resulting in vigorous volcanic activity with a high output of carbon dioxide to the air, and a rapid reaction of the rocks with the ocean waters, producing copious hydrogen gas. A weaker, orange sun shone in a probably orange-tinted sky.

Archean
3.7–2.5 billion years ago

The Archean saw early bacterial life forms and the birth of Gaia. The atmosphere after life appeared would have been mainly nitrogen, with carbon dioxide and methane at between 0.1 and 1 per cent. Oxygen would have been present as a trace gas, rapidly used up by the reducing compounds of the Archean environment. A pale pink sky may have arched over a brown ocean.

Proterozoic
2.5–0.7 billion years ago

The transition from the Archean to the Proterozoic was marked by a switch from a reducing to an oxidizing atmosphere. The Earth was still populated by bacteria; in the anoxic sediments, the Archean or prokaryotic bacteria would have lived on, but in the mildly oxidizing surface environments, more complicated eukaryotic cells would have developed (see Chapter 5). With oxygen rising, the sky may have been pale blue over a greenish sea.

Phanerozoic
0.7 billion years ago, up to the present

The Phanerozoic is the time of the plants and animals, when oxygen rose in abundance to 21 per cent. Carbon dioxide continued to be pumped down by living organisms to its present low level of 0.03 per cent, to compensate for the Sun's increasing luminosity. The landscape we know begins to emerge, with its verdant land life, bluer skies and seas, cloudiness, and occasional forest fires.

The period of the Earth from the beginning until just before life began is well named the Hadean (see opposite) because of its fiery violence. Not only were planetesimals constantly colliding, making craters hundreds of kilometres in diameter, but in the hot early Earth volcanism must have been many times more active than now. The Hadean lasted nearly a billion years and during that time the Earth cooled and evolved gases that became its first lasting atmosphere. The gases of the air then would have been carbon dioxide, water vapour, nitrogen, carbon monoxide, and small amounts of hydrogen. Just what the composition was is less important than the fact that it was the right mixture to keep the surface warm in spite of the fact that the Sun was 25 per cent less luminous than now. The Hadean was so long ago that no rocks remain to provide scientific evidence about the conditions of those violent times. But it is probable that there was plenty of water, and enough hydrogen being produced to prevent free oxygen appearing.

The early Earth was evolving entirely according to the laws of physics and chemistry. Gases coming from volcanoes were being removed by chemical reactions. Carbon dioxide, from the volcanoes, was reacting in the presence of water with basalt rock to form the carbonates of the elements sodium, potassium, calcium, magnesium, and iron. The same reaction process removed water. Elements such as iron reacted with water, sequestering the oxygen and releasing hydrogen gas. The hydrogen sweeping up through the waters and the atmosphere would have caused a steady loss of water. Hydrogen gas breaks down to hydrogen atoms at the outer limits of the atmosphere and, being too light to be held by the Earth's gravity, escapes to space. Had this process continued, in one or two billion years the Earth, like Mars and Venus, would have lost all of its water and become irreversibly a dead planet. It is near certain that in these early times both Mars and Venus had water in abundance.

Here we glimpse the first signs of Gaia. Water on the Earth enabled life, but without life the Earth now would be dry. Life retained water in two important ways. First, certain microorganisms were able to gain energy by using hydrogen to make hydrogen sulphide, and these captured and retained the hydrogen produced during the reaction of water with basalt rocks on the sea floor. Second, the very act of photosynthesis releases free oxygen as a by-product, and in the reducing atmosphere of the Archean some of this oxygen would have combined with the escaping hydrogen gas to produce water. So you see, without life and oxygen, the oceans of the Earth would in a few billion years have vanished forever, as the hydrogen that once was part of the ocean water escaped to the vacuum of space and merged with the restless stream of atoms forever blowing from the Sun.

But I have digressed. We need to think about the start of life at the end of the turbulent Hadean period. At some time toward the end of this period the Earth became cool and stable enough for the chemical reactions to occur on the surface that paved the way for the start of life. Somewhere, somehow, compounds of carbon and of other elements formed that were stable enough to react further and form ensembles of compounds and structures. These ensembles would be recognizable as characteristic of living organisms.

Scientists spend much time on research and speculation into the origins of life. Some think it happened by a chance combination of the molecules and structures of the early Earth; others favour an origin out in space on pieces of cometary debris that eventually fell on the Earth and seeded it. Still others see a distant origin with the seeds of life posted here purposefully by others or even by God.

As a scientist I am naturally interested in how life started; but I am also content to accept that it did, without knowing how. To me it is not important whether life originated by chance, floated in on a piece of cometary debris, or was placed here by others, even including God. I am more concerned about what happened once Gaia came into existence – and I am fairly sure that Gaia must have been born *after* the start of life, not before or at the same time.

My reasons for thinking this way are these. As the post-Hadean Earth cooled and sorted out its chemistry, there would have been a period when the chemistry and the climate of the Earth were favourable for life. But this favourable state could not have persisted indefinitely. The balance between the output of carbon dioxide from volcanoes and its removal by reaction with the basic rocks would have needed to stay constant, otherwise the gaseous greenhouse of this gas would have increased and overheated the Earth, or decreased and frozen it. (For a detailed explanation of the greenhouse effect, see pages 137–8). In the longer-term the loss of water as it reacted with the rocks and made hydrogen would have dried the planet out. Before the Gaia hypothesis, it was commonly thought that the conditions on Earth were just right for life by a fortunate accident. Although this was true for the start of life, it is now apparent that these fortunate conditions at the origin were transient. Without Gaia the physical and chemical evolution of the Earth would soon have moved to a state inhospitable for life. It was the evolution of the tightly coupled system, life and its environment, that sustained the small range of temperatures and chemical compositions that was and still is favourable for the persistence of life.

Once life did start it would, slowly at first but eventually on a larger scale, have affected the Earth's environment. Scientists do not know what the first organisms were. They may have been bacteria that used the freely available energy of sunlight, or they may have been bacteria that gained energy by fermenting the organic debris left over from earlier unsuccessful attempts at life. We can be sure though that soon after life started, photosynthesis – the harnessing of the Sun's energy to make simple sugars (see p. 55) – became the primary source of energy for life.

CO_2

Basalt

Hydrogen loss

During the Archean, there would have been a continuous production of hydrogen gas from the reaction of oxides in basalt rock with carbon dioxide and water. Water would have been split (see diagram) releasing hydrogen to the atmosphere, and locking the oxygen into the various carbonates of sodium, potassium, calcium, magnesium, and iron. Two important consequences would have arisen from this reaction. First, the maintenance of an oxygen-free atmosphere and surface, providing a favourable environment for the accumulation of life chemicals. And second, the loss of hydrogen to space. Hydrogen atoms are extremely light, and the Earth's gravitational field is not strong enough to prevent their escape.

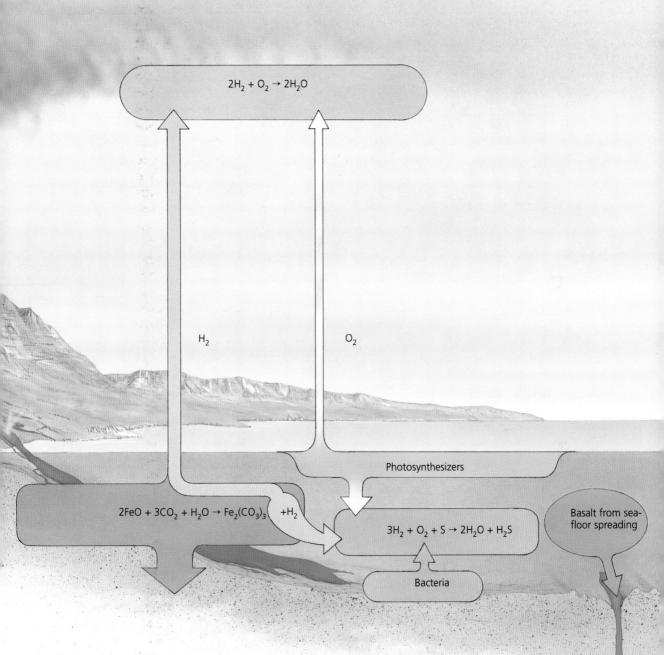

$$2H_2 + O_2 \rightarrow 2H_2O$$

H_2

O_2

Photosynthesizers

$$2FeO + 3CO_2 + H_2O \rightarrow Fe_2(CO_3)_3 \quad +H_2$$

$$3H_2 + O_2 + S \rightarrow 2H_2O + H_2S$$

Basalt from sea-floor spreading

Bacteria

How life has kept the planet moist

If hydrogen had continued to escape to space, then within a few billion years, Earth might well have lost all its water and become an arid, lifeless planet like Mars or Venus.

Fortunately, however, life intervened. First, by adding oxygen to the environment, as a by-product of photosynthesis (and of carbon burial). Some of this oxygen would undoubtedly have combined with free hydrogen in the reducing atmosphere of the Archean to form water, preventing its loss to space. Second, the free hydrogen produced on the ocean floor (see diagram) would have been used by certain bacteria to gain energy by making hydrogen sulphide, and the hydrogen thus would have been retained. The presence of life in the Archean saved our planet from a dry and dusty death.

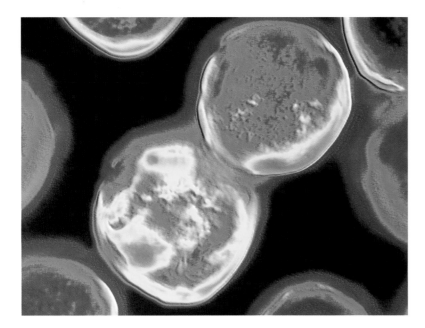

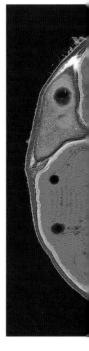

1 A false colour light micrograph of the thermophilic bacteria Archaeoglobus fulgidus which reside near submarine hot thermal vents in temperatures in the region of 83 degrees celsius. It is able to reduce sulphates for anaerobic growth, and has been identified as a metabolic "bridge" between the rest of the thermophilic genera (which reduce elemental suphur) and the methanogenic bacteria. This is one of the Archaebacteria whose predecessors populated the earth during the Archean.

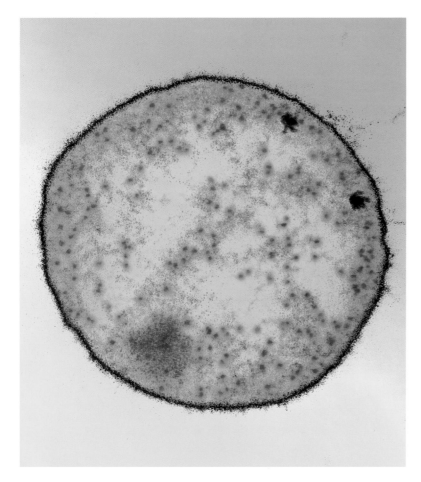

2 This transmission electron micrograph (TEM) of a section through another archaebacterium, a cold-loving species called Methanococcoides burtonii. It can survive temperatures as low as minus two to five celsius. As a methanogenic bacterium it is able to form methane from carbon dioxide and hydrogen. The scattered genetic material can be clearly seen within the cell.

Sunlight is the only abundant, ubiquitous, and almost everlasting source of energy. Which particular chemical route was used by the early photosynthesizers is a topic of speculation. They may have first used the easy sulphur cycle, splitting the weakly bound hydrogen sulphide, or they may have tackled the more difficult task of splitting oxygen from water. In either case, to manufacture sugars they would have had to use carbon, and carbon dioxide as the source. No other carbon compound was as plentiful or in continuous supply from volcanoes. The growth of photosynthesizers would have reduced the atmospheric abundance of this gas and consequently the protective warmth of the gaseous greenhouse. A biosphere one hundredth as active as the one today could have removed all of the carbon dioxide in a few million years.

Once life had substantially affected the chemistry of the Earth it had also altered its own environment. At the start of the Archean period, about 3.8 billion years ago, the Sun was 25 per cent less luminous than now. At that ancient time, the removal of the carbon dioxide greenhouse was as hazardous as is our addition to it now. If the growth and spread of the first photosynthesizers had gone on unchecked or uncompensated, they would have reduced the greenhouse until their environment froze, and then eked out a thin existence on an arctic world, until either they died out, or conditions improved.

I think that this course was checked by organisms called fermenters; they are the scavengers of the bacterial world who make a living from the small store of chemical energy left in the excreted products or dead bodies of the primary producers. They may have existed before the photosynthesizers appeared, or could have evolved to occupy the vast niche of litter once life was fully established. Fermenters gain a small chemical energy return by converting organic matter to methane, ethanol, or other simple molecules. Yeast does this with sugar, but complex organisms such as yeasts did not evolve for over a billion years. It was achieved in the Archean by the methanogens, fermenters that converted organic matter into methane and carbon dioxide (see opposite). Some of this carbon dioxide would have escaped to the air. Moreover, in the oxygen-free environment of the Archean oceans, methane would have been much less reactive than carbon dioxide, and nearly all of this gas, which also has a greenhouse effect, would have reached the air. Together these two gases would have kept the greenhouse intact and so sustained a favourable climate for early life (see also Chapter 7).

The birth of Gaia came when the evolution of those simple bacteria according to Darwinian natural selection and the evolution of the planetary surface environment and atmosphere ceased to be two separate and independent processes. Once life began to change the atmosphere then natural selection ensured that the change

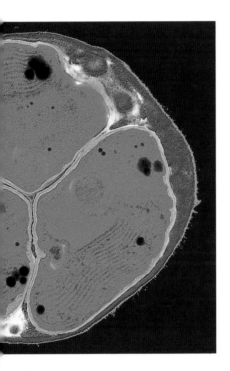

3 The eubacteria, which developed much later than the archaebacteria during the Proterozoic, have a more complex evolved internal structure. This TEM of Cyanidium caldarium, a eubacterium, clearly shows, as well as other organelles, three large chloroplasts which have evolved within this species as a result of endosybiosis. It is a photosynthesizer, as the chloroplasts show, and is an acidophile which can survive acidity around pH 0, very rare in a living organism. Its optimum growth temperature is around 50 degrees celsius.

could only be toward a favourable environment. The self-regulation of the Earth's climate and chemistry then became a natural and inevitable consequence, or as the philosophers would say, it became an "emergent" property of the system (see p. 88). Emergent because the entire system when working had properties absent from a mere collection of all the component parts.

Gaia's childhood diseases

The birth of an animal is a risky time in its life. It is vulnerable as a newborn organism. It has not yet taken charge of itself and of its homeostatic mechanisms. In the period after life had appeared on Earth and before the system Gaia existed, life was like an early fetus with the Earth as its womb. Able to grow and flourish because, for a brief precious period, the Earth's climate and composition were moving through the region ideal for life, womb-like and comfortable. But unlike the womb, the Earth's environment was changing, moving in such a way that if life failed, the planet itself would develop lifelessly to a state like that of Mars and Venus now, and become forever after uninhabitable for life.

Life's infancy is a vulnerable time, and Mars and Venus may never have survived that hazardous period. Unfortunately for our understanding, it was all so long ago that hardly any direct evidence remains from which to draw conclusions about the pathology of infant planetary organisms.

The same kind of uncertainties exist about Gaia's childhood. The Archean period lasted over a billion years after life began, and was, as with human childhood, a period with little or no interest in, or use for, sex. Scientists are sure that there must have been many stresses and disaster during those times. Many large planetesimals (some more than 10 kilometres in diameter) must have hit the Earth and each would probably have destroyed more than half the life that existed before the impact. Volcanism is likely to have been more violent than now. The heat that drives the tectonic forces of the Earth comes mostly from the radioactivity of the elements that go to make our planet. As the atoms of radioactive elements such as potassium, uranium, and thorium decay, they let go, little by little, of that same vast store of energy that, if instantly released, creates a nuclear explosion. It is this energy that heats the rocks. In the Archean the heat was about three times greater than it is now because there was then three times as much radioactivity. There may have been metabolic catastrophes as species of bacteria evolved to occupy some niche and in so doing exerted a subtle disadvantageous effect upon the remote environment. Imagine, for instance, a bacterial ecosystem that made a business of chlorine chemistry. There is plenty of chlorine in the sea and not as much energy is needed to make oxygen from water. It would have been an unpopular act to have tried to turn the ocean into a chlorinated swimming bath. Chlorine reacts with water to

Photosynthesizers

The details of life's origins are ineffable, but soon after the first forms of life appeared on Earth, organisms must have learned to use the premium energy of sunlight to grow and to sustain themselves. The first photosynthesizing bacteria used sunlight in a complex series of steps to take carbon from carbon dioxide from the air and ocean and use it to build their bodies. The present system of photosynthesis using chlorophyll was an early step. Oxygen was excreted as a by-product. The early Earth environment was reactive toward oxygen and there was no accumulation of this gas as now.

Methanogens

A world with photosynthesizers only would be unstable. They would soon have locked up in their bodies most of the available carbon. Their removal of carbon dioxide would have so weakened the greenhouse that the world would have frozen, and life ceased. This never happened. There coexisted with the photosynthesizers simple fermenters, the methanogens. These organisms processed the organic matter made by the photosynthesizers and returned carbon to the air as a mixture of methane and carbon dioxide, restoring the greenhouse. A bonus from the presence of methane was the creation of an "organic smog" layer in the upper atmosphere, shielding the surface from ultraviolet radiation, much as the ozone layer does now (see p. 140).

Consumers

Limited to isolated pockets near the surface, where enough oxygen would have been produced by the photosynthesizers to support them, early consumers would have lived on the organic products of the photosynthesizers.

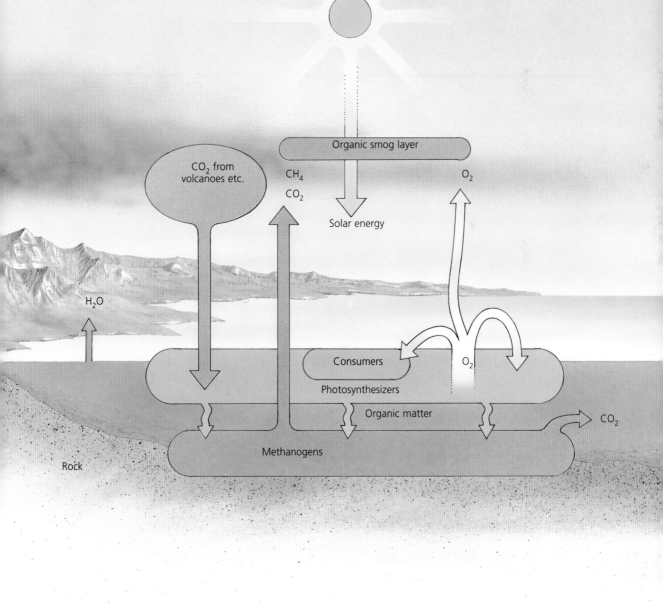

CASE HISTORY Nuclear Dermatitis

Gaia was afflicted at puberty by an interesting, perhaps to us alarming, but, as events have shown, not a serious condition. About 1.8 billion years ago (during the Proterozoic) in a region that is now Gabon in Africa, the newly dominant element oxygen began to affect the composition of water flowing through a marshy region. The water that had fallen as rain, perhaps on mountains or hills nearby, had weathered the rocks and dissolved the usual elements calcium, magnesium, silica, iron and sulphur. But in addition the water contained significant quantities of uranium in the form of the oxidized uranyl ion. Now the oxidized uranyl ion is soluble in water, whereas the uranous ion, which existed previously in the oxygen-free environment of the Archean, is not. The presence of dissolved uranium in the water during the Proterozoic period was to lead to an interesting pathology.

Quite apart from its radioactivity, uranium is a toxic element. If you swallow a gram of uranyl acetate you will die not by radiation from the element, but by ordinary chemical poisoning. The organisms of the marsh through which the uranium-enriched water began to flow evolved to survive in its presence by fixing the element in the walls of their cells rather as if they were making a shell. This is an old trick of living organisms. The abundant element calcium is needed, indeed essential for life, but strangely it is also more toxic than cyanide. Abundances of calcium over a few parts per million are lethal. Cells neatly solve the calcium problem by making sure that any excess is stored as inert calcium carbonate in shell or as calcium phosphate in bone. Not surprisingly, having learned the trick they did the same thing with the uranium. We humans do a similar thing. Calcium is put in our bones where it serves as a structure for our bodies as well as a safe store of a dangerous but essential element. We also if we encounter uranium or plutonium put it in our bones out of harm's way chemically. With us, of course, a highly radioactive element like plutonium is then put close to the vulnerable cells of the bone marrow. Sadly the chemical wisdom of the body is what kills the plutonium eater.

To return to our organisms in the Proterozoic marsh, they fixed themselves shells of uranium salts. They were probably quite undisturbed by the mild radioactivity of the uranium, as microorganisms are far less sensitive to radiation than large animals. In any event individual microorganisms do not in general live long enough to be damaged by radiation. What they did not know was that the uranium, while not a radioactive hazard, was in those times enriched in the fissionable isotope ^{235}U.

Uranium now contains only 0.7 per cent of ^{235}U, the rest is the more stable isotope ^{238}U. Two billion years ago the proportion of ^{235}U was 5.6 per cent. At the beginning of the Archean it would have been more than 18 per cent. The greater the proportion of the fissionable isotope ^{235}U, the easier it is to make a nuclear reactor. Indeed, at even higher proportions a bomb is possible. Fortunately, only the highly skilled can today make and assemble the materials of a nuclear reactor. Back in the Proterozoic, dumb, insensate microorganisms could do it and they did.

The uranium collected by the organisms was wonderfully pure and free of neutron-absorbing elements that can easily poison a reactor, and when enough had collected a natural nuclear reactor began operation and ran for a million years or so. So far as Gaia was concerned this was a very minor illness, perhaps no more than a cosmetic disturbance. A nuclear pimple upon the fair face of the Earth. There may have been other reactors elsewhere – if so they have not yet been discovered, or they may have been eroded away or buried deep in the Earth.

I sometimes wonder what it would have been like had the Earth been oxidizing soon after life began. Uranium nearly four billion years ago would have been so enriched in the fissionable isotope ^{235}U that its aggregation by organisms would have led to nuclear firework displays of a marvellous kind. The illness would have been much more severe but I suspect that Gaia would have coped and possibly even turned the phenomenon to use.

Yellowcake

The photograph, left, shows a flask of "yellowcake", a form of uranium oxide. It was mined in Gabon, Africa, where marsh organisms long ago fixed the uranium in their cell walls, so creating a nuclear reactor.

Fissionable uranium

Uranium-235 is the fissionable isotope of uranium. It is extracted from the oxide and used as a fuel in nuclear reactors and as the explosive in nuclear weapons. The 4.5kg "button" shown above was once worth about US$200,000.

A nuclear pimple in Gabon

The rich uranium mines of Gabon, above, owe their wealth to uranium-fixing microorganisms in the Proterozoic marshes.

form hydrochloric and hypochlorous acids. In sea water there is enough bicarbonate to neutralize the hydrochloric acid, but some hypochlorous acid would remain free – and this acid can penetrate cell membranes, and is destructive to all organic matter. Perhaps this is why few organisms use chlorine in their biochemistry and why methyl chloride, although detectable in the air, is a trace component only (see p. 122).

For numerous brief periods in the long history of the Earth, the never-ceasing experiments by the various forms of life must have led to a catalogue of planetary pathologies. (One such example has just been given on pages 86–7, that of nuclear dermatitis, a relatively minor disturbance.) But in Gaia such metabolic monkey-wrenching is self-limiting and cannot grow so as to destroy the system. The rules of Gaia are that organisms that adversely affect their environment do not survive long. We humans would do well to remember this.

Emergent properties of systems

SYSTEM	EMERGENT PROPERTIES
Electric iron or oven with thermostat	Thermostasis
TV or radio receiver	Homeostasis of display when signal varies
Automatic pilot	Homeostasis of velocity, direction, and altitude
Bacterium	Chemostasis on internal environment, including regulation of pH, pE, and electrolytic balance. Morphostasis (keeping form constant)
Animal	Chemostasis and thermostasis
Gaia	Homeorhesis

CHAPTER FIVE

Biochemistry and the cell

Surprisingly, most of our planet is fit for living organisms. It is neither too hot nor too cold, the sea is not too salt, nor is it too acid or alkaline. Oxygen, absent in the Archean, is now abundant, but there is ample room in the muds and sediments for those descendants of the Archean, the anaerobes who cannot live in the presence of oxygen. The Earth is amazingly right for life, and has been so for nearly four billion years.

Adaptation by life is not an adequate enough explanation for this fortunate state of affairs. Powerful geophysical and geochemical forces are always working to change the planetary environment away from that narrow band of conditions favourable for living cells. The real question about Gaia is: has everything been so near perfect here on Earth for so long by a happy accident? Or has the Gaian system of life and its environment continuously evolved so as always to sustain an environment favourable for life? And if so, how? How have living organisms, in their chemical exchanges with the world around them, created and sustained a favourable environment? This chapter, and the next, attempt to answer these questions, tracing the links between the intricate chemistry of life and the closely coupled evolution of living organisms and the planetary environment.

Getting to know Gaia involves seeing and feeling it as the huge and ancient entity that it is, a top-down view from wherever one looks. Equally it requires a feeling for the microscopic levels of existence, for the myriad small, brief lives that contribute to Gaia. The illustration on page 91 shows what a vast range of sizes is covered by life. The scale of size is expressed in ten-fold steps, each being ten times larger than the one before. Such a scale is the only one that can encompass all the sizes of life on the same page.

I shall start by considering the genes, and then the cells, the units of life, and so move on through time and space to the larger animals and plants, to show how life and the environment could have evolved together in the close-coupled system of Gaia. But before discussing the genes, let's take a closer look at the science most appropriate to this chapter – biochemistry – and consider its strengths and weaknesses, and its significance for the understanding of planetary medicine and Gaia.

The divided sciences

The Victorians were careless when they allowed science to divide and become an array of sectarian expertises. Each newly separated science soon developed its own argot and gang of professors who ruled, from the cloistered turrets of their universities, over sharply bounded fiefs. Internal strife ensured further divisions and so it was that the great barony of chemistry split into inorganic chemistry, the science of that part of matter that was not and never had been alive, and organic chemistry, which concerned the special and intricate chemicals of living organisms.

Early chemists long believed in a vital force that made chemicals from living organisms wholly different from inorganic chemicals. This belief was the justification for the separation of organic and inorganic chemistry. It implied that organic chemicals such as sugar or fat could be analysed but never synthesized in the laboratory. The German chemist Friedrich Wöhler (1800–82) in 1828 was the first to question the faith in vitalism. He took what to him was an unequivocal inorganic chemical, ammonium cyanate, and showed that when heated it transmuted into urea. Until Wöhler did his experiment, urea was categorized as an organic chemical, something made only by living organisms. Soon chemists found that they could make other organic chemicals, but it was not until the chemist Hermann Kolbe (1818–84) synthesized the familiar organic chemical, acetic acid, that vitalism was at last rejected. Chemists went on to synthesize the intricate molecular structures of cholesterol, vitamins, and even proteins. Now it looks as if everything existing in living organisms can be synthesized from "inorganic" materials.

The recognition that the old division of chemistry into organic and inorganic was an artificial one did not, at the time, lead to the wider recognition that the chemistry of life and the chemistry of the material environment are part of one continuum, of Gaia. And, of course, it made not a jot of difference to the chemist's self-imposed apartheid. Inorganic chemists continued to work with minerals, metals, and strong corrosive chemicals. Organic chemists rarely consorted with their inorganic counterparts, and soon found themselves under pressure to divide yet again as the chemical industry of the 19th century burgeoned. Industry now recruited so many organic chemists that the subject's biological background faded. The chief concern of organic chemists became the industrial application of carbon chemistry for commercial synthetic products. Those still interested in the chemistry of life became the nucleus for a new science that soon crystallized as biochemistry.

Sizes of life

Like the "Ages of life" illustration on page 75, the scale of size of living objects is here expressed in thousand-fold steps, each step being one thousand times larger than the one before. Such a scale is the only one that can encompass all sizes of life on the same page. If we take the smallest virus as the smallest manifestation of life, and Gaia as the largest, then the range of linear dimensions goes from 0.1 micrometres to 13 megametres. The range of masses goes from 1.0 femtograms to 10^{32} grams. The largest life form, Gaia, is 10^{47} times greater in mass and 10^{15} times greater in linear dimension than the smallest, the virus.

Raindrops

Molecules

Atoms

10^{-}

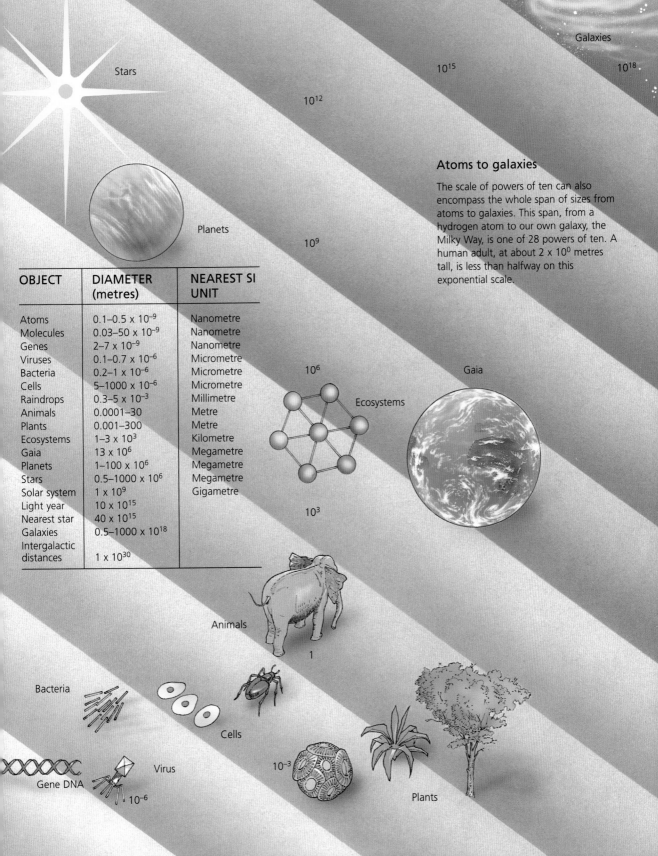

Galaxies

Stars

10^{15}

10^{18}

10^{12}

Planets

10^9

Atoms to galaxies

The scale of powers of ten can also encompass the whole span of sizes from atoms to galaxies. This span, from a hydrogen atom to our own galaxy, the Milky Way, is one of 28 powers of ten. A human adult, at about 2×10^0 metres tall, is less than halfway on this exponential scale.

OBJECT	DIAMETER (metres)	NEAREST SI UNIT
Atoms	$0.1–0.5 \times 10^{-9}$	Nanometre
Molecules	$0.03–50 \times 10^{-9}$	Nanometre
Genes	$2–7 \times 10^{-9}$	Nanometre
Viruses	$0.1–0.7 \times 10^{-6}$	Micrometre
Bacteria	$0.2–1 \times 10^{-6}$	Micrometre
Cells	$5–1000 \times 10^{-6}$	Micrometre
Raindrops	$0.3–5 \times 10^{-3}$	Millimetre
Animals	$0.0001–30$	Metre
Plants	$0.001–300$	Metre
Ecosystems	$1–3 \times 10^3$	Kilometre
Gaia	13×10^6	Megametre
Planets	$1–100 \times 10^6$	Megametre
Stars	$0.5–1000 \times 10^6$	Megametre
Solar system	1×10^9	Gigametre
Light year	10×10^{15}	
Nearest star	40×10^{15}	
Galaxies	$0.5–1000 \times 10^{18}$	
Intergalactic distances	1×10^{30}	

10^6

Ecosystems

Gaia

10^3

Animals

1

Bacteria

Cells

10^{-3}

Virus

Gene DNA

10^{-6}

Plants

Biochemistry is about the fundamental molecules of life. From its lowly beginnings in the pathology departments of hospitals, where biochemists spent much of their lives analysing blood, urine, and faeces, it has now blossomed into the life sciences equivalent of particle physics. Together they form the last level of reduction. While physicists disembowel the quarks to discover the meaning of the universe, biochemistry probes and dissects deoxyribonucleic acid (DNA) molecules – the stuff of genes – to decode the messages they carry.

Molecular biology itself started with the joint work of both chemists and physicists in the laboratories of the Medical Research Council in Cambridge, England. It was here that they discovered the magical double helix of DNA. My friend, Sir John Gray, who was secretary of the Council for many years, reminded me that it was a discovery requiring so much scientific ecumenism that it could never have come from any of the separated sciences alone. Once discovered it soon became the property of the biochemists. They were the scientists with the laboratories and the mental equipment needed to work with DNA, and unravel the language of the gene.

Gaia and the selfish gene

Richard Dawkins, in the first of his books, *The Selfish Gene,* has done more than any other writer to express in plain language that we proud humans, like all other living things, are no more than the outward expression of our genes. The idea of a gene, a fundamental unit of genetic inheritance that determines the nature of an organism, was born out of the work of Gregor Mendel (1822–84), the monk who wondered about the inheritance of colour in peas. Geneticists soon realized that, in a world where Darwin's natural selection ruled, the ultimate entity selected was not the organism, but the gene. Our bodies are to our genes as is the cathedral to its architect. We exist so that the genes can replicate and continue their existence.

But what is a gene? A gene is like an atom. Neither gene nor atom is an ultimate particle: each is made from smaller units; and both live long and are consistent. The gene is a coherent message written in the language of the DNA molecules. But a gene is more than just a message. A gene is a recipe for a part or property of a living organism: anything in any organism from a bacterium to a whale. More still, a gene is a programme that orders and oversees the work of construction so that everything needed is made and assembled and put together at its due time and place.

In his second book, *The Extended Phenotype,* Richards Dawkins tries to show that Gaia could never exist because there is no way for the genes to express themselves on a planetary scale. I will now show how the genes *can* speak to the planet and how the environment of the planet affects the survival of the cell and therefore that of the genes it bears – and vice versa.

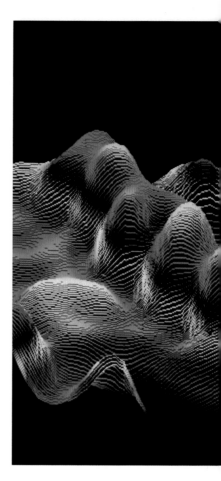

DNA under the electron microscope

Magnified 1,600,000 times under an electron microscope the ridges of the DNA double helix can be seen: they correspond to the orange/yellow peaks to the left of centre in the photograph (above). The genetic message contained in the DNA molecule, however, is difficult or impossible to comprehend. It is like a recipe stated in a paragraph of a cookbook written in ancient Egyptian. You can see it but not understand what it means.

The chromosome

Chromosomes are thread-like structures contained within the nucleus or control centre of the cell. When the cell is about to divide, they become shorter and thicker, and can be seen to be made up of two parallel strands, called chromatids.

Along the length of each chromosome is a series of chemical structures called genes, which are the basic units of inheritance.

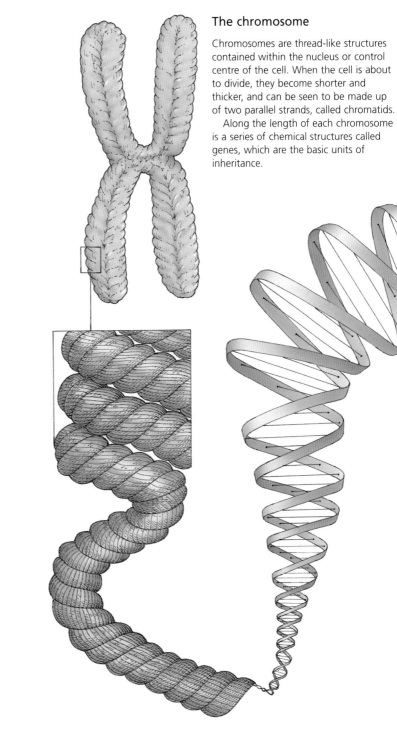

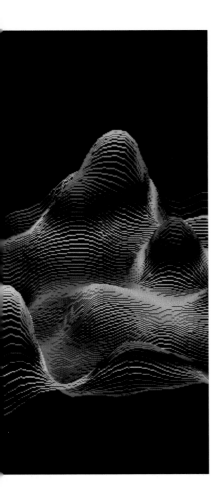

Unravelling the DNA

A chromosome consists of a protein framework, around which is coiled a long DNA molecule. The genes are actually made up of sections of this long DNA molecule.

A strand of DNA consists of a long chain of nucleotides. Each nucleotide is made up of a simple sugar, deoxyribose, combined with a phosphate group, and an organic base, which may be one of four types. The nucleotides are linked in chains by their phosphate groups. Molecules of DNA comprise two of these long strands, running parallel, and held together by bonds between the bases, to give a ladder-like structure. This is twisted into the characteristic spiral shape – the famous double helix.

Breaking the genetic code

The secret of the genetic code lies in the sequence of the four organic bases of the nucleotides – adenine, guanine, cytosine, or thymine – along the molecule of DNA. It is this code that instructs the cells to assemble amino acids to make particular proteins. Each sequence of three bases provides the code for one amino acid. The diagram below shows a segment of the twin-stranded DNA, with its phosphate groups, sugars, and paired bases.

The four bases are like letters of an alphabet. Each letter can be used to form words, sentences – even books – to carry the message and programme of life.

The replication of DNA

When a cell divides to produce another cell, the chromosomes are copied exactly so that the new cell inherits a full set of chromosomes. The copying process is known as replication, because the DNA molecule in each chromosome, prior to cell division, makes an exact replica of itself. The twin strands part, and each acts like a key for assembly of a new twin. The nucleotides along its length hook up with others available in the medium, pairing opposite bases – cytosine with guanine, adenine with thymine – and so replicate the complete sequence that forms the coded genetic information on the double helix.

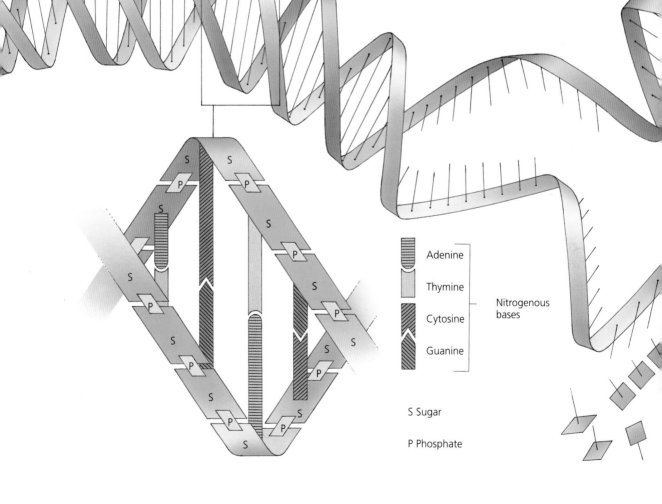

Adenine
Thymine
Cytosine — Nitrogenous bases
Guanine

S Sugar

P Phosphate

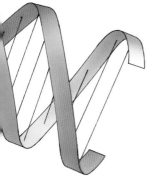

Gene, cell, and planet

The key connection that links biochemistry and the gene with Gaia and planetary medicine lies in the properties of some simpler molecules that are part of living cells. DNA is exciting and important stuff, but it depends for action on other structural and working molecules without which the DNA could not replicate. Most important are the special but ubiquitous structures, the cell membranes – structures bound together by forces as weak as those that bind a soap bubble.

Physicists acknowledge that the microscopic properties of the atoms and their parts are linked with the large-scale properties of the universe. The evolution of the universe, the galaxies, the stars, and their planets all depend upon the properties of their atoms. The physical force that determines the lifespan of atoms (and therefore of planets, stars, and galaxies) is the "weak" nuclear force. Called weak because the other nuclear forces are so much stronger, it is the force that determines the rate of radioactive decay of an element. Solid matter can only exist because the weak force binding its atoms together is strong enough to stop its radioactive disintegration.

In a similar way the membranes of living cells are held together by weak intermolecular forces known as Van der Waal's forces. Were the planetary environment to change, these weak forces might no longer be able to hold together the molecules of the membranes, and life itself might vanish.

Molecular biologists have yet to stand back from their reductionist view of life to see that the exciting properties of DNA are linked through the weak forces of biochemistry (that maintain the cell membranes of all living organisms) with the large-scale properties of the Earth.

The vital cell membrane

Cells could never function without membranes. There are inner membranes around the various organelles inside the cell, in addition to the outer membrane separating the cell from its environment. Both inner and outer membranes have many functions. First, to secure all the important structures and molecules of the cell against the universal tendency for dispersion. Then to sustain an internal environment at just the right salinity, acidity, and degree of oxidation. For these last functions the membranes include pumps that regulate the flow of ions and molecules. They are much more than just passive walls. Often they can manage the bulk transfer of food or waste particles as large as other cells, and of course they can accommodate the division of the cell when it reproduces.

The cell membrane is essential and irreplaceable, yet it is the most fragile and vulnerable part of an organism. To break the bonds between atoms, high temperatures, such as cooking in an oven, are needed; but the tenuous, soap-bubble-like forces holding together the cell membrane are weak enough to be broken by no more than even hot or cold weather. Membranes are also sensitive to changes in salinity, acidity, and the structure and form of the water in which they are immersed.

What happens to the membrane when things go wrong? Imagine a house made from blocks of ice. It can be done, as the igloo of the Inuit proves. Ice is a weak material and a high-rise building of ice would flow under its own weight into an unseemly puddle. A rise in temperature to above 0°C would be as disastrous, as would exposure to salt. The cell membrane is like the shell of the igloo. The forces holding the membrane lipids together are not strong. Most cells will disintegrate at temperatures above 50°C and at a salinity 30 per cent higher than that of sea water.

The survival of a cell and its membranes (and therefore of the genes it expresses) requires it to be in an environment tolerable for its membranes – it must be able to live with the prevalent tempera-ture, salinity, acidity, redox potential, and water availability. Life on Earth is composed of living cells. So for life to persist, the physical and chemical conditions of the environment must remain within that narrow band suitable for the cell membranes. And life has persisted, for nearly four billion years. For this to occur by chance is, to say the least, improbable. That is why I say that the system of life and its environment, Gaia, must be responsible.

All organisms change their environment through their life chem-istry. And, as Chapter 3 showed through the Daisyworld geophysio-logical models, the pressures of selection can act in favour of those organisms with a tendency to create changes that improve the envi-ronment. If this improvement favours the survival of the cells and their membranes, it also, of course, favours the survival of their progeny, and therefore of their genes. Thus the weak forces that bind the cell membrane affect all life on Earth, and link the gene, through the natural selection advantage of improving the environment of the cell, with the global environment. Richard Dawkins and other disci-ples of Darwin are right to reject any notion of a genetic *control* of the environment. Nonetheless, the busy activity of the genes does massively and continuously modify the material environment of the Earth. These modifications feed back to exert selective pressure on the next generation of organisms.

But what of those highly specialized cells such as the thermophilic bacteria that can live in boiling water, and other specialist organisms that can exist in saturated brine. Couldn't all living organisms adapt to these extremes? The answer is no.

Cells living at high temperatures or in brine are less efficient and more vulnerable than the cells of regular forms of life, and often depend for their nutrients on mainstream life. I sometimes compare my eccentric existence as a scientist living in the country with that of a halophile bacterium (see p. 100). I do not farm, hunt, or fish for my food, nor do I hew wood to burn to keep me warm. My exis-

The structure of cell membranes

Cell membranes are complex arrangements. They are built from structural materials, proteins and polysaccharides, which establish the framework – somewhat like the girders of a steel-frame building, but elastic and flexible – and lipids that make the membrane impermeable. In the walls are a set of biochemical pumps to move water, food, electrolytes, and waste products to and from the cell interior. There are also gates controlled by enzymes for entry or exit of large items such as other cells; mechanisms to allow motion; and, most important, a factory for producing the transient materials, the lipids and other products that need replacing to maintain the membrane's intact existence. Most cell membranes have outer coats of polysaccharide, too, as protection against adverse environmental change. The diagram, right, shows all these structures.

The most vulnerable part of the membrane is the lipid layer – a bimolecular sandwich structure in which phospholipids are arranged with their electrically charged, water-soluble "heads" outward and their water-insoluble "tails" inward (like rows of tadpoles joined tail to tail, see diagram).

The lipids are held together by the same weak forces as those that sustain a soap bubble, forces whose strength is only a twentieth that of the bonds holding the atoms of carbon and hydrogen that go to make the lipid molecule. The vulnerable lipid layer is vitally important to the membrane and the cell. A hole less than a 1000th the area of the membrane of, for example, a red blood cell would allow its contents to escape in less than a second. Dead cells from which the contents have escaped by membrane failure are still visible under a microscope, though the lipid layer has gone. Such cells, called "ghosts", show that the cell membrane has structure as well as the lipid double layer.

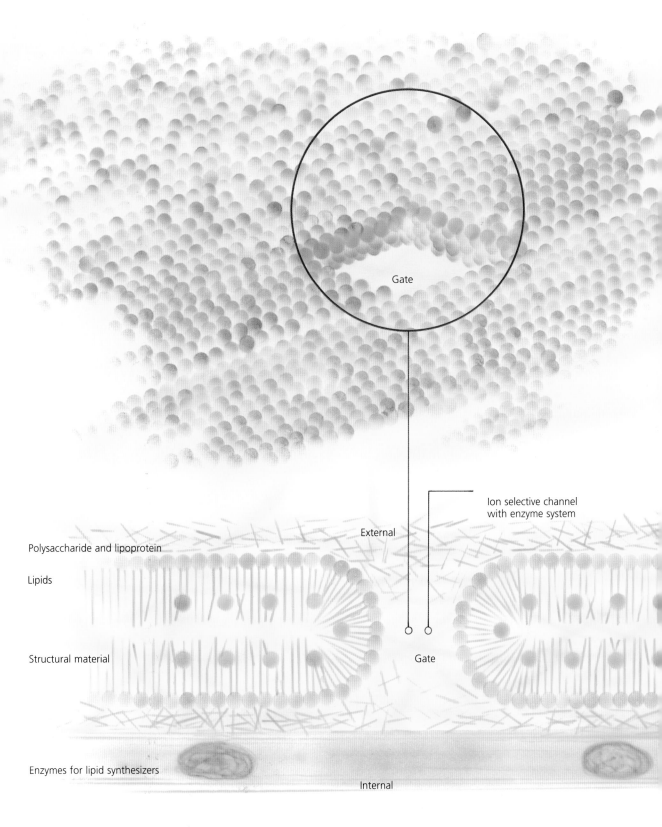

Gate

Ion selective channel
with enzyme system

External

Polysaccharide and lipoprotein

Lipids

Structural material

Gate

Enzymes for lipid synthesizers

Internal

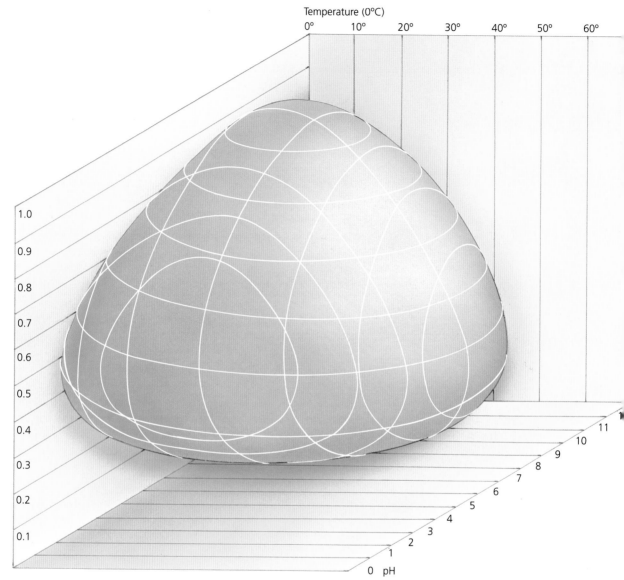

A tolerable environment

The essence of understanding Gaia is to recognize the environmental constraints that shape the form and behaviour of organisms. The constraints within which a living cell can survive include factors such as temperature, humidity, acidity (pH), salinity, ionic strength, and redox potential. A diagram of the window for life defined by *all* environmental constraints would form a hypervolume – a space that cannot be mapped on a page. The diagram above maps just three critical constraints: temperature, ionic strength, and pH (acidity).

Organisms require a limited range of conditions for these factors – temperatures from 0°C to 50°C, pH from 3 to 9, and ionic strength from 0.1 to 0.9. The ability of a cell to tolerate fluctuations in its environment depends upon the ability of its membranes to withstand disruption, and the forces that hold a cell membrane together are only weak (p. 95). Organisms that live in extreme environments must have evolved special strategies to enable their cells to survive. For example, plants and animals living in the desert generally have good

water conservation; organisms such as marine algae have devised means of converting salts to harmless substances; and so on. Remarkably, however, conditions in the Earth's environment do generally meet the narrow constraints for cell survival. This fact is the strongest evidence for Gaia, the system of life and its environment.

Viruses

The smallest truly living thing is an *Aphragmabacterium*, which is about one fifth of a micron in diameter. But what about viruses, which are even smaller? Aren't these alive? Perhaps, but I prefer to think of viruses as no more than pieces of the software of life. They are just a set of instructions expressed in the common genetic code and wrapped up in a convenient protein package. They are like computer floppy disks bearing a coded message. Neither the virus, nor the disk, is by itself alive. The disk needs a computer for its expression, and the virus, a living cell. The viral "message" enters the living cell and commandeers its function, causing it to act out its instructions. With the virus there is always the message "Having done all this, make a copy of the entire set of instructions and pass it on". Mischievous hackers place similar "virus" messages on computer disks. But neither the virus nor the disk can do anything by themselves.

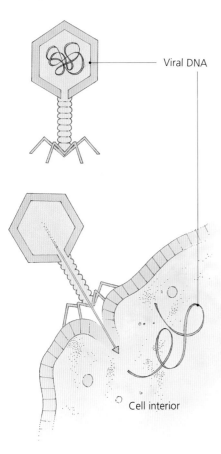

Viral DNA

Cell interior

tence is possible because the rest of society sustains me within a favourable economic climate. Like the halophile, I could not survive alone. We both need the support of a rich and vigorous society.

The evolution of cells

So closely coupled is the evolution of living organisms with the evolution of their environment that together they constitute a single evolutionary process in which life has, quite literally, fashioned the environment to suit itself. And in this remarkable story, it is the small and lowly organisms that have played the starring role.

The smallest things that are truly alive are the prokaryotic microorganisms, casually classified as bacteria. These tiny organisms are complete: they can exchange matter and energy with their environment; maintain a constant size and composition, and reproduce. They come in all shapes and sizes, from the spheres of the cocci to the rods of the bacilli and the thin spirals of the spirochetes. Some swim free, propelled by slender fibres that thrash the fluid of their medium like tiny oars, while others are less mobile: the sarcina bunched in cubes like the atoms of a crystal; the staphylococci growing like bunches of grapes; the streptococci in chains.

To most of us, bacteria are intimately connected with disease. We think of them as tiny malign germs that lurk in dirt. Indeed the science of bacteriology grew up in the pathology labs of hospitals. As always with human knowledge, our view is mainly restricted to the small segment that is a human concern, the pathogenic bacteria. Yet bacteria are, and always have been, the most important living things. Their ceaseless activity in the soil, the sediments, animals and plants, is essential for the continued existence of Gaia, indeed of life itself.

Bacteria have made and will make up, so long as organic life persists, the greater part of life on Earth. They play the largest role in sustaining Gaia. In the soil and in the sediments beneath the waters of the lakes and the oceans they are so numerous that they make up a significant fraction of the total mass. They exist even in dark ecosystems deep below the Earth's surface, gaining energy by reacting buried carbon with the oxygen chemically stored in the calcium sulphate of gypsum deposits. They can be found in abundance at the depths of the oceans manipulating the chemistry of plumes of hot water emerging from the sea-floor spreading processes (see Chapter 2, pp. 42–3). In short, bacteria are everywhere throughout the Earth, even, as I will soon explain, as captive slaves within our cells and in those of other animals and plants.

So efficient is communication among the bacteria that for many purposes the whole world of bacteria can be thought of as a single organ. It always has a structure such that the photosynthesizers and consumers are on the outside in the light and oxygen, and the fermenters below in the dark. To the fermenters oxygen is a poison.

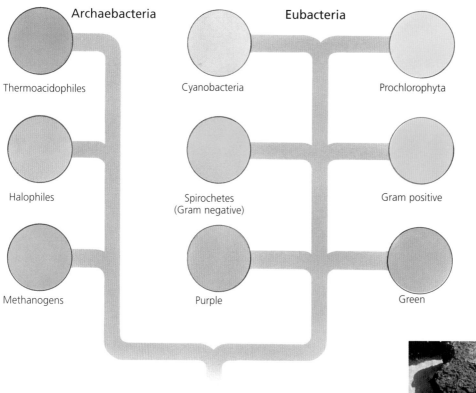

Archaebacteria

Thermoacidophiles

Halophiles

Methanogens

Eubacteria

Cyanobacteria

Prochlorophyta

Spirochetes
(Gram negative)

Gram positive

Purple

Green

Types of bacteria

All bacteria share a common feature: they have a prokaryotic cell structure. That is to say, their cell structure is much simpler than that of other organisms. A typical bacterium or prokaryotic microorganism consists of a cell filled with cytoplasm, but with no membrane-bound nucleus, and no organelles (specific bodies within the cell). Instead it has a single chromosome coiled at its centre. Bacteria exchange genetic information freely with the environment, and divide by simple fission.

There are two broad classes of bacteria: the ancient Archaebacteria, and the far more numerous Eubacteria, or "true" bacteria. Archaebacteria typically reflect their evolutionary origins in the early environments of life. They include the methanogens, which occupy anaerobic (oxygen-free) sediments; the halophiles, which thrive in salt-rich locations; and the heat- and acid-loving thermoacidophiles of sulphur springs and volcanic deep-sea vents. The more recent Eubacteria can be grouped into at least the six divisions shown in the diagram (and possibly more). They include the chlorophyll–containing photosynthesizers, such as the blue-green cyanobacteria and the prochlorophyta; the chemically distinct gram positive and gram negative (or spirochete) bacteria, and the purple and green bacteria, which may have begun as anaerobic photosynthesizers, but later evolved aerobic heterotrophic species. Research by the American scientist, Woese, suggests that the life story and family relationships of bacteria may be different and more complex than I have just said.

Stromatolytes

These stromatolytes, now growing at Shark Bay, Western Australia, are like those that existed between 2 and 3 billion years ago. They are rocky, cushion-like masses of cyanobacteria (blue-green algae) which are preserved in the calcium carbonate they secreted (see p.130). These structures were abundant during the Archean, when the cyanobacteria were the most advanced form of life on Earth. Stromatolytes are still being formed in coastal waters such as Shark Bay.

If you want to know more of this world of microorganisms you should read the works of my close friend and colleague Lynn Margulis. Read especially *Microcosmos* that she wrote with her son, Dorion Sagan, or her own enchanting books *A Garden of Microbial Delights* and *Symbiotic Planet*.

The Archean, which ran from about 3.8 to 2.5 billion years ago, was the great age of the single-celled, prokaryotic organisms, the bacteria. Then only traces of oxygen, probably less than one part per million, were present in the air. In those times, consumers would not have flourished. There may have been a few consumers among the bacteria of the Archean, but they would have been limited to the immediate vicinity of dense growths of the photosynthesizing bacteria (cyano- or blue-green bacteria). In such places oxygen (the by-product of photosynthesis) may have been abundant enough locally for consumers to exist.

I must not leave you with the impression that in the bacterial period of life on Earth, there were no large organisms. Communities of bacteria including photosynthesizers, consumers, and fermenters grew together with the rocky material they were transforming into structures called stromatolytes, some of which were as large as houses. Such communities or ecosystems are still to be found today (see photo, left).

From complex cells to sex

During the Proterozoic period, which followed the Archean, oxygen began to dominate the chemistry of the air and the nature of life on Earth changed profoundly (see Chapter 6. pp. 111–13). Whole ranges of habitats must then have become available to consumers, as must large numbers of new species of organisms to occupy them. I suspect though that some of these new species would have descended from the Archean consumers, and not have appeared suddenly once oxygen became generally abundant.

Lynn Margulis is distinguished for her support of the endosymbiosis hypothesis, a view that I find extremely compelling. In this hypothesis the cells of all the more recent life on Earth are made up from assemblages of smaller organisms that once were free-living. A cell, for example, in the leaf of a plant contains within it small green bacteria-sized bodies. These are called chloroplasts and enable plants to use sunlight to produce organic matter and oxygen from the raw materials carbon dioxide and water. According to Lynn, these bodies are the descendants of one free-living cyanobacteria. Our own cells contain similar small bodies to the chloroplasts of the plants. These are called mitochondria, and they perform the reverse energy transaction by recombining oxygen and organic matter to provide the power that we need to live. Mitochondria and chloroplasts are genetically quite different from the genetic material of the cell nucleus. (There are, for example, distinct genetic diseases of the mitochondria.) I find this convincing evidence for the view that these organelles were once separate and free organisms.

Endosymbiosis began when, sometime early in the history of life, consumers ingested other bacteria, or bacteria were invaded intracellularly by other bacteria. These events were, of course, usually fatal to the victims, but once in a while, instead of the death of one or other of the pair, a truce was called and they lived together in symbiosis, each benefiting from the other's presence. An ingested cyanobacterium, instead of dying, continued to photosynthesize and feed not only itself but the cell that had eaten it. Endosymbiosis led to the emergence in the Proterozoic of a new kind of cell, the eukaryote. Larger and more complex than a bacterial cell, eukaryotes are the cells from which all of the larger manifestations of life, animals, plants, fungi, and protoctists (unicellular organisms), have evolved.

Our cells contain many organelles that may once have had a separate existence, with different genes. I have long wondered if the mammalian egg is large compared with the spermatozoon because it has to carry as passengers all the organelles and their genetic material as well as the genes of the nucleus. Bacteria (the ancestors of the organelles) reproduce by fission and exchange genetic material through plasmids, as these processes could all go on in the egg as it develops.

The need for sex

There are several competing explanations of the need for sex but the one I find compelling is as follows: in the Archean, simple bacteria exchanged genetic information freely. The whole environment was pervaded with bits and pieces of genetic molecules. Any organism that had forgotten its instructions, or that needed the code to deal with a problem new to it, could pluck the missing information from its surroundings. This was a fine and efficient state of affairs and allowed bacteria to manage the planet for over a billion years unaided. Then, in the Proterozoic, the presence of oxygen and consumers made possible, and necessary, a new and interesting invention: sex.

The rise of free oxygen enabled a rapid turnover of the synthesized organic matter by consumers, and it was the consumer ingestion of cell by cell that led to the evolution of eukaryotes.

A cell containing several organelles might find it difficult to gather genetic material from the medium fast enough to maintain health in the face of loss by mutation (which is also increased by oxygen). Some organelles, like the mitochondria and protoplasts, could go on being bacteria, and gather in their plasmids and divide at will. Others, such as the more organized cell nucleus, would need something more subtle. We do not know how or when it happened, but the first eukaryotic cells to transfer genetic material in neat packages

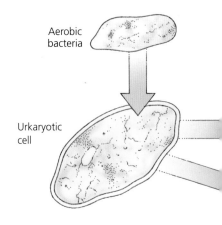

Archean cells

Aerobic bacteria

Urkaryotic cell

The endosymbiosis hypothesis

A eukaryotic cell is much more complex than a prokaryotic bacterium: it has an enclosed nucleus and a full complement of organelles, and is typical of the vast majority of the cells that make up modern species – animals, plants, protoctists, and fungi. So how, during the course of evolution, did eukaryotic cells arise from the prokaryotes?

Endosymbiosis, or "living in partnership" is increasingly accepted as the explanation. Long ago, during the Proterozoic period, a prokaryotic cell ingested by another cell did not die, but survived to live in partnership with its host. Such host cells, the precursors of true eukaryotes, are often called urkaryotes. The ingested cell was probably an aerobic purple bacterium, a heterotroph (consumer). According to the endosymbiosis hypotheses, it is from this ancient partnership that all modern cells have evolved.

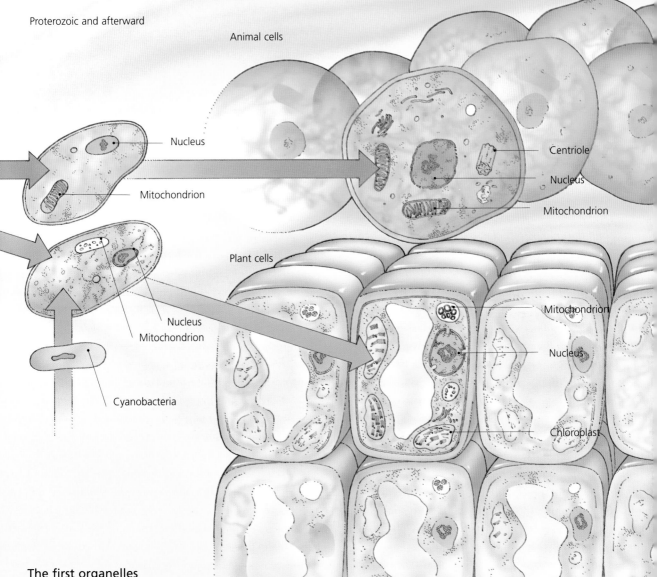

Proterozoic and afterward

Animal cells

Nucleus

Mitochondrion

Centriole

Nucleus

Mitochondrion

Plant cells

Nucleus
Mitochondrion

Cyanobacteria

Mitochondrion

Nucleus

Chloroplast

The first organelles

The ingested bacteria living symbiotically within their host cells had their own genetic material and divided independently. They were the precursors of the mitochondria of modern cells. Some of the symbiont's genetic sequences may have been borrowed by the host, to form part of the evolving cell nucleus. Later, another type of bacterium, a photosynthesizer – probably one of the blue-green algae, or cyanobacteria – also took up symbiotic residence within some of the evolving cells. It was the precursor of the chloroplast. Cells with this new organelle started the evolution of plant life.

Eukaryotic plant and animal cells

All eukaryotic cells have a separate, membrane-walled nucleus, within which lie multiple thread-like chromosomes, wound round a protein core, and enclosed in an inner membrane. They also contain many organelles. Mitochondria are found in both animal and plant cells. But chloroplasts are found in plant cells only, and are responsible for their unique photosynthetic ability. The endosymbiotic origins of the mitochondria are apparent in several ways. They have, like their prokaryotic ancestors, a single circular chromosome, no nuclear membrane,

separate genetic instructions, and their own outer cell wall, which differs in structure from that of the eukaryotic cell itself. Modern cells are highly complex and vary hugely in size, function, and organelle content. It is possible that some others of these organelles, such as centrioles, may also have an endosymbiotic history. Even now the process of co-operation continues: some present-day photosynthetic and heterotrophic bacteria live symbiotically inside host cells, sharing food and energy.

The advantages of being multicellular

For billions of years the oceans have provided (and continue to provide) a far more stable medium for the development of life than the land surfaces. Oceans are not subject to wide fluctuations in temperature: nutrients tend to be more evenly dispersed (though concentrations vary with physical mixing and plankton activity); and water, as a denser medium than air, provides buoyancy and support to simple organisms lacking a rigid structure.

If the land surfaces were to be effectively colonized by life, then the development of multicellular organisms was a prerequisite. The advantage of communal living was that it enabled certain groups of cells to specialize in particular roles. Over time, multicellular organisms thus evolved various strategies for coping with the harsher conditions of the land environment. For example, they developed specialized cells forming a skin or cuticle, to reduce water loss by evaporation, and some evolved clusters of cells acting as sense organs, to facilitate rapid response to environmental change. Structural problems of life on land were solved by more rigid cells in plants, and skeletal frameworks in animals. And throughout, organisms were not only adapting to the new environment, but, by their ceaseless activity, adapting and shaping the environment itself.

Completing the pattern

Sexual reproduction provides an extra safeguard against loss or damage of genetic information between generations, due to mutation and duplication errors. The incomplete information carried separately by the sperm and by the egg can be likened to separate copies of a notice, damaged by stray bullets on a battlefield. The gunfire is unlikely to have damaged the same words on each of the two copies. There is a good chance therefore of reconstructing the complete notice from two damaged copies. As the diagram below shows, damaged areas of the genetic coding sequences on a chromosome can be restored when the two parental sets are combined. In nature, even if there is a common mistake carried by both male and female genetic elements, natural selection among the progeny will ensure that only those altered or incomplete copies that are fitter than the rest will survive.

Combined

had a large advantage over their rivals who gathered lost information simply by browsing.

Sex, for us, is the most delightful emotional and sensual entertainment that life can provide; it is also part of our way of reproducing. Less often mentioned is the importance of sex as a way of restoring missing information. A significant piece of genetic instruction might be missing, or incomplete, in the genetic material of the sperm or of the egg. On conception, the two information sets combine and the incomplete parts are most unlikely to be the same for both the sperm and the egg. From two flawed lists, one pristine set of instructions is made and, if not, natural selection disposes of the error.

Multicellular organisms

At some unknown period during the Proterozoic, certainly earlier than a billion years ago, communities of eukaryotic cells grew together in synchrony and the first multicellular organisms evolved.

A strong constraint on the size of the first multicellular organisms must have been the need for an efficient system to transfer nutrients to, and wastes from, the organisms at the centre of the community. Single microscopic cells can take in oxygen and nutrients and excrete wastes simply by diffusion through their membranes to and from the surroundings. But a spherical community of cells, larger than a few millimetres in diameter, would have difficulties in sustaining the supply of oxygen, and removal of carbon dioxide, to and from its interior. This problem would be much greater if this communal organism used energy for motion, or to keep warm. The advantages of community living can be had, however, without paying the penalty of asphyxiation or starvation. The evolution of two-dimensional flat organisms or of open-mesh three-dimensional sponges illustrated two of the possible strategies. A more sophisticated solution to this problem came with the development of circulatory systems, with a network of channels and a means to sustain a flow of blood or plasma through them. For us and other animals these are the heart, arteries, veins, and lymph channels. For insects and other invertebrates, simpler systems suffice.

Important among the advantages given by sheer bulk was the ability easily to colonize the land surfaces. I don't doubt that algal mats and other communities of individual cells were present on the land from the earliest times. But the land was a harsher environment than the oceans; they must have been thinly spread, and certainly did not flourish like modern plants and animals. The success of the larger organisms comes mainly from their greater ability to resist desiccation and cope with large variations of temperature, even freezing. Just as oxygen and nutrients can easily diffuse into a single cell suspended in the ocean, so water can leave the same cell rapidly if it is cast up on the shore. A large tree or animal can hold on to its water long enough to survive between wet and dry spells.

Once communities of cells came together as multicellular organisms, there was a greater need for oxygen, and greater means for increasing its supply through the burial of larger quantities of carbon (see Chapter 6). Oxygen in the air may have begun to arise in abundance as organisms that needed it evolved, possibly in the region of a billion years ago.

Throughout this long evolutionary process, the constraints on cells of living in environments of the right temperature, moisture, acidity, salinity, and so on, would have given a steady, selective pressure. Natural selection would have favoured organisms whose collective effect was to improve the environment, and/or capitalize on the environmental changes (such as the rise of oxygen) caused by other organisms. The balance of carbon dioxide, oxygen, methane, nitrogen, and trace gases in the air, the salinity of the oceans, the rate of formation of soil – all are so closely woven into the evolutionary patterns of living organisms as to form a single, changing tapestry. The role of life in these larger, planet-scale phenomena is the subject of the next chapter.

CHAPTER SIX

Metabolism and planetary biochemistry

To define life merely as something that replicates and can correct the errors of replication by natural selection among the progeny is not enough. Such a definition is too broad and could apply to a computer game. Life as we know it is also characterized by metabolism. Something that takes in materials and free energy from the environment, makes chemical transactions, and excretes waste products and low-grade energy as heat. The metabolic (or biochemical) definition of life includes plants, animals, and, as will become apparent, Gaia. In this chapter we look at the planet-wide metabolic processes of Gaia – of life and the environment – and discover how they have shaped (and continue to shape) the air, oceans, and land.

Metabolism and the atmospheric gases

Sunlight is the only significant energy source available on Earth. A small flow of heat comes form the Earth's hot interior, but this is less than 1 per cent of the heat received from the Sun. The quality of the Sun's radiation is more important than the quantity. Energy in the form of mere heat (that is, long wavelength infrared radiations) is not potent enough for plants to conduct their business of splitting carbon dioxide into oxygen and food. It is the white light of the Sun that enables plants to use sunlight to sever the carbon-oxygen bond of carbon dioxide. If the Sun were just 500°C instead of blazing white at 5700°C, and if the Earth were in a closer orbit and just as warm as now, no life would be possible. The potential energy of light coming from a cool Sun is insufficient to allow plants to break the chemical bond between carbon and oxygen. Even with the white light of our Sun, plants still need to add the potential energy of several photons to conduct their metabolism.

Plants store the energy they gather from sunlight as chemical potential energy, in much the same way as batteries, when charged, store electrical energy. Burning the plant in the oxygen it has made releases the stored energy as heat. Obviously plants alone could not form a planetary ecosystem. They would either run out of carbon dioxide, or conflagrate as the excess of oxygen they had made inflamed the least spark of fire from lightning or volcanism. Two other main ecosystems keep the balance. First the decomposers (or methanogens) that take the debris of dead plants and convert it to carbon dioxide and methane. And second, the consumers that gain energy from the slow internal combustion of food and oxygen.

The exchange of matter and energy between these great divisions of life was what 19th-century scientists called the balance of nature. Many biologists today seem to think that this alone explains the levels of the two great metabolic gases – carbon dioxide and oxygen – in the air. This view is wrong. The picture of the world it gives is like that of a ship with the pumps connected merely to recirculate the bilge water within it, rather than to pump it out. As the water leaked in, the ship would soon sink.

Carbon dioxide

On the real Earth carbon dioxide is always leaking in from volcanoes. If there were not some means for its removal, it would steadily rise. The biota recycle vast quantities of carbon dioxide, food, and oxygen between the producers and the consumers, but these exchanges are almost in balance; the imbalance although small is vital for sustaining the levels of oxygen and carbon dioxide (see page 112). Nonetheless carbon dioxide does not rise, because it is always leaking out of the system, by burial either as carbonate rock or as carbon itself.

So what is this "leak" that thus determines the level of carbon dioxide in the atmosphere? In short, it is rock weathering – or, to be more precise, the rate of reaction of carbon dioxide with calcium silicate rocks. Basalt rock, the solid dark exudate of volcanoes, is rich in calcium silicate, and when it is immersed in rainwater saturated with carbon dioxide, it slowly dissolves. The products of this reaction are a water solution of calcium bicarbonate and silicic acid. This solution percolates through the ground waters into the rivers and to the oceans. The calcium carbonate and the silica end up on the ocean floor as sediments and eventually are buried, to form the source of limestone (calcium carbonate) rock strata.

Until the 1990s, geochemists maintained that the presence of life has had no effect on this set of reactions. It is simple chemistry that determines the level of carbon dioxide in the atmosphere. Life or Gaia, they said, is not needed to explain what happens. But I disagreed. With my friends Michael Whitfield and Andrew Watson, I published papers in the early 1980s proposing that the presence of organisms in the soil greatly enhances the rate of rock weathering. We stated that by their growth, plants pump carbon dioxide from the air into the soil, citing as proof the observed 10- to 40-fold enrichment of carbon dioxide in the air spaces of the soil (as compared to the air above). When a tree dies, for example, most of its mass is eventually oxidized through the action of decomposers and

Volcanic activity

The carbon dioxide cycle

Carbon dioxide is the key metabolic gas of Gaia, influencing climate, plant growth, and oxygen production. It cycles constantly through the system from its source, volcanic output, to its final sink, burial as limestone (calcium carbonate). The level of carbon dioxide in the air, (currently 0.03 per cent), depends on the balance between the rates at which it leaks in and is pumped out. Plants by their growth break up surface rocks and draw down carbon dioxide into the soil. There, dissolved in rainwater, it reacts with basalt rocks to form calcium bicarbonate, which is washed down to the sea and used by the microscopic marine life to form shells. The ocean algae also pump down carbon dioxide from the air. When the microflora die, their shells rain down to the ocean floor, to form sediments of limestone and chalk.

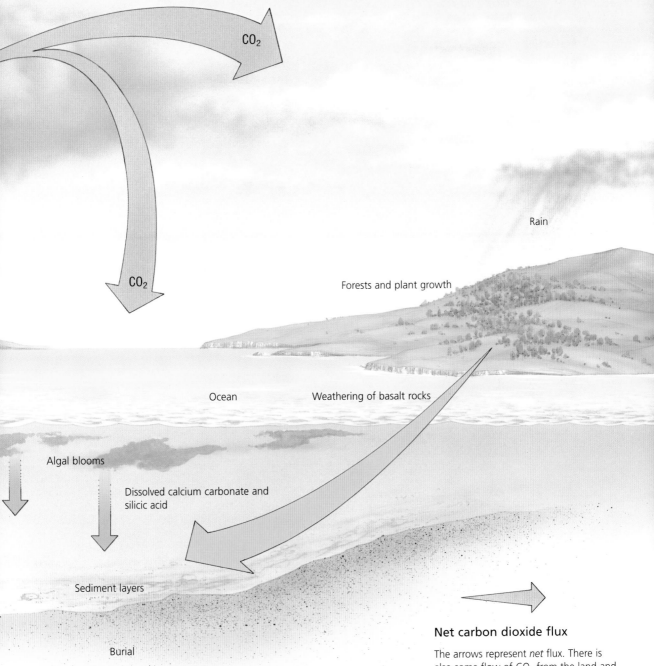

CO$_2$

CO$_2$

Rain

Forests and plant growth

Ocean

Weathering of basalt rocks

Algal blooms

Dissolved calcium carbonate and silicic acid

Sediment layers

Burial

Net carbon dioxide flux

The arrows represent *net* flux. There is also some flow of CO$_2$ from the land and sea back to the air, but it is less than the flux from the air to land and sea.

The meadows of the sea

The sunlit upper layers of the ocean are populated by countless millions of free-floating microscopic plants, and are the meadows of the sea. The microflora pump down carbon dioxide from the air during their rapid spring growth, and use dissolved salts to form their delicately structured shells. Diverse and beautiful, the flora of the euphotic zone include coccolithophores, which make their shells from calcium carbonate, and diatoms and radiolaria, which use silica. The rain of intricately patterned dead shells or "tests" forms, over millions of years, the sea-floor sediments of limestone and silicate strata.

Radiolaria
(*after Haeckel*)

Coccolithophores

converted to carbon dioxide. The site of much of its conversion is in the soil right next to the calcium silicate rock in the presence of water. In 1989, the scientists T Volk and D W Schwartzman, in an article in *Nature*, showed that in fact the weathering of basalt rock was a thousand times faster in the presence of organisms than with sterile rock.

In the ocean, too, life is involved. Some algae (the diatoms) take the silicic acid to make their skeletons, and other algae (the coccolithophores) take the calcium bicarbonate to make skeletons of calcium carbonate. Ocean algae also pump carbon dioxide from the air. Scientists are just beginning to realize how blooms of algae on the ocean surface first fix carbon dioxide in diatoms, and then after a subsequent great bloom of another algae, the coccolithophores, convey the now solidified carbon dioxide to the sea floor in their heavy carbonate shells. So you see, from the weathering by bacteria of the exposed rock of a continental mountain, to the deposition of algal shells as limestone on the ocean floor, there is an unceasing intervention by organisms in the great carbon dioxide cycle.

The old reductionist way of doing science caused scientists to consider the cycle of carbon dioxide as if it were complete in itself. Geophysiologists know that it is not and that it is inseparably linked with the other cycles of the Earth, as if Earth were one organism. If carbon dioxide varies, so does climate (see Chapter 7), and so does the rate of plant growth and the production of oxygen. The fixing of carbon dioxide is so fundamental to life that no living process can be entirely independent of it.

Oxygen

If the level of carbon dioxide is a property of the Earth as an organism, is the level of oxygen also? Oxygen, like carbon dioxide, has varied greatly during the history of the Earth. The illustration overleaf shows the variations of oxygen, methane, and carbon dioxide during the history of life on Earth.

Scientists are nearly certain that the Archean period of the Earth was anoxic. That is, the abundance of free oxygen in the environment was very low or zero. I think that there would have been a trace of free oxygen in the air – perhaps one part per million, like methane now. Photosynthetic bacteria were abundant in the Archean and some oxygen may have leaked from them into the atmosphere.

The abundance of free oxygen in the atmosphere depends upon the balance between its production and removal. In the Archean, this balance was low or negative, like the bank balance of a heavy spender. A good supply from the source (the photosynthetic bacteria), but always the expenditure was larger than the income and the account overdrawn. The expenditure was the reaction of the oxygen with organic matter (simultaneously produced by photosynthesis), with methane, and with reducing substances such as ferrous iron and sulphur compounds. These reducing substances were continuously released by volcanic processes and by the activity of microorganisms. (Photosynthesis alone can never sustain a net production of oxygen. It merely separates the oxygen and hydrogen elements of water and the carbon and oxygen elements of carbon dioxide – and these can and do recombine to form water and carbon dioxide again. A net increase of oxygen comes either from the escape of some of the separated hydrogen to space, or by the burial of carbonaceous material synthesized by organisms.)

Early in the Archean oxygen was scarce or absent, and as a result there was no significant population of consumers grazing the photo-synthesizers and returning carbon dioxide to the atmosphere. Instead, there were the methanogens, early bacteria capable of existing only in the absence of oxygen, which lived by decomposing the organic matter of the photosynthesizing bacteria and converting the carbon in it to carbon dioxide and methane.

One or two drops of a strong acid, such as hydrochloric, are enough to dominate the chemistry of a glass of water and make it acid – it does not require 20 per cent or more. In a similar way methane, produced by the methanogens, dominated the chemistry of the Archean atmos-phere. There was less than 1 per cent of methane but this was enough to control the atmospheric chemistry. (This theoretical prediction about the Archean atmosphere came from my geophysiological computer model of Gaia's early climate – see page 137).

During the latter part of the Archean, when the flux of oxygen removers (iron and sulphur from tectonic and volcanic sources) began to decline, there was a gradual growth of oxic organisms or consumers at the surface, where there would have been enough oxygen produced by the photosynthesizers to support them. The consumers would have spread to cover most of the oceans as the oxygen-scavenging compounds of the sea were all used up. The decline of tectonic and volcanic activity was a probable consequence of the diminishing stock of primeval radioactive elements whose decay supplies the Earth's interior heat.

Geological evidence suggests that, throughout the Archean, a small but constant proportion of the carbon turnover of the photo-synthetic bacteria would have been buried. The rate was not that different from today, and would have led to the steady addition of oxygen to the atmosphere, but this would have been countered by its removal by reducing substances. The end of the Archean was marked by a sudden change from a methane-dominated to an oxygen-dominated atmosphere, when the gain of oxygen from carbon burial became more than that of its removal by reducing compounds. The critical point was reached when there were two molecules of oxygen for each molecule of methane.

$$CH_4 + 2O_2 \rightarrow CO_2 + 2H_2O$$

Methane + oxygen → carbon dioxide + water

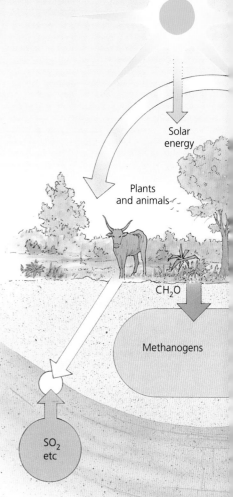

Oxygen balance sheet

	Credit	Debit
Plants	100.1	
Animals	–	99
Methanogens	–	1.0
Rocks	–	0.1
	100.1	100.1

Oxygen, methane, and carbon

Plants photosynthesize and convert carbon dioxide and water into oxygen and organic matter, of composition approximately CH_2O. Animals and microorganisms consume most of this, using up the oxygen made by plants and returning carbon dioxide to the air. About 1 per cent of the organic matter is buried deep in the soil, where methanogens convert it to carbon dioxide and methane.

The methane escapes to the air, where it reacts with the remaining 1 per cent of oxygen to form water and carbon dioxide.

A small proportion (about 0.1 per cent) of the buried organic matter escapes digestion by the methanogens. The carbon is buried deep in the sedimentary rocks and the equivalent oxygen is left free. Thus it is this small amount of buried carbon that accounts for the oxygen in the air. All the rest of the oxygen made by the plants is used up by animals and microorganisms, by reaction with methane and with rocks and gases during volcanic activity, and by weathering.

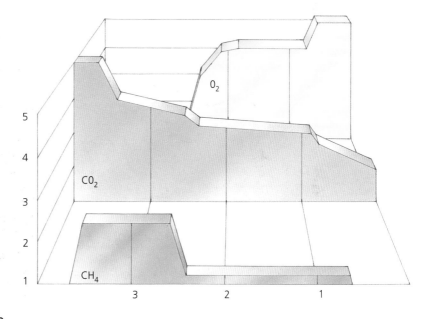

The evolution of the atmosphere

The involvement of life in Earth's atmospheric composition is illustrated by the diagram above right. It shows the geophysiologist's view of the evolution of the atmosphere during Gaia's long life. The abundance of gases, shown on the vertical scale, is expressed as parts per million (ppm). (The scale is logarithmic, that is, in powers of ten: 1 means 10 ppm, and 5 means 100,000 ppm.)

The horizontal axis shows the timescale expressed as eons before the present. The diagram is based both on geological evidence and on my computer models of Gaia (see Chapter 7).

It shows the progressive stages of decline of carbon dioxide levels from an abundance of between 10 and 30 per cent ($1–3 \times 10^5$ ppm) before the birth of Gaia to the low level of 0.03 per cent (3×10^2 ppm) now, due to the intervention of life in the weathering process and carbon cycle. It also shows the early appearance of methane, generated by the methanogens (fermenters) of the Archean, and its subsequent decline

alongside the rise of free oxygen. This complete change of state from a reducing to an oxidizing atmosphere heralded the rise of the consumers. You can see how oxygen climbed to around 1 per cent (10^4 ppm) during the Proterozoic and remained there until, around the time of the evolution of larger land-based life, with greater carbon burial, it rose toward its current 21 per cent (over 2×10^5 ppm).

Oxygen and fire

Why did oxygen remain at 21 per cent, and not rise higher? I think the answer is fire. The correlation between oxygen abundance and flammability is steep. Below 15 per cent, nothing will burn; above 25 per cent combustion is instant and awesome fires would rage, destroying all forests. Charcoal layers in the geological record show that oxygen has long been above 15 per cent, and remains of ancient forests show that it has not exceeded 25 per cent. But how could the oxygen–fire relationship in practice act as a Gaian regulatory mechanism? An answer could lie in the fire ecology of forests: certain species, the conifers and eucalypts, do include fire in their evolutionary strategy; others do not. As with the dark and light daisies in Daisyworld, the competition for space between the trees could provide a feedback control on oxygen and fire (see Case History on p. 129).

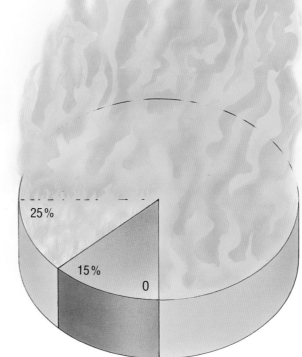

Percentage O_2 in atmosphere

Oxygen – polluter or eutrophicator?

In the early stages of the Gaia hypothesis, Lynn Margulis and I often talked of the rise of free oxygen in the atmosphere as if it were a disaster. We called it the first and greatest air pollution incident, where vast ranges of species of organisms were killed off by oxygen poisoning.

I now think that the opposite is true. Oxygen is poisonous, it is mutagenic and probably carcinogenic, and it thus sets a limit to lifespans. But its presence also opens abundant new opportunities for organisms. At the end of the Archean, the appearance of a little free oxygen would have worked wonders for those early ecosystems. Oxygen was good for two reasons: first, its presence enables consumers to recycle the organic matter of the photosynthesizers much more rapidly than could be done by the methanogens; and second, oxygen would have changed the environmental

chemistry. The oxidation of atmospheric nitrogen to nitrates would have increased, as would the weathering of many rocks, particularly on land surfaces. This would have made available nutrients that were previously scarce, and so allowed an increase in the abundance of life.

Geophysiological models of the change (see Chapter 7, p. 137) suggest that even the denizens of the anoxic regions would have benefited. This is simply because oxygen is, in a sense, a eutrophicator – that is, by increasing nutrient levels it encourages surface "blooms" of algae, and these deny light and oxygen to life below, which soon dies. The increased supply of dead plants in its presence would have enriched the methanogens until the consumers took over this role.

Below this point, despite the actual abundances, there was a methane-dominated atmosphere, but once past it oxygen ruled the Earth and has done so ever since.

After its appearance at the end of the Archean, oxygen probably remained for some time at a stable abundance of about 1 per cent by volume in the air. Too much oxygen is toxic, too little slows the rate of biochemistry. The Proterozoic period that followed, from 2.5 to 0.7 billion years ago, is poorly understood. Scientists know that for the most part it was a world of microorganisms like the Archean. Unicellular life was hardly vigorous enough to bury the larger quantities of carbon needed to sustain a high level of oxygen by counteracting its rapid removal by reaction with the rocks. I think that oxygen did not increase much above 1 per cent until the evolution of large plants and animals.

Large animals, particularly if they move, need plenty of oxygen in the air, at least 10 per cent. More oxygen could have become available when large, particularly land-based plants evolved. The carbonaceous matter of large plants is more likely to be buried in the sediments than that of large microorganisms. Atmospheric oxygen must have risen rapidly at some time after the land was extensively colonized and as more and more organic matter was buried. What set the limit at 21 per cent was, I think, flammability.

Fire is a splendid indicator of oxygen. The relationship between flammability and oxygen abundance is steep. A 1 per cent change of oxygen increases or decreases the chance of ignition by nearly 100 per cent. Below 15 per cent oxygen fires will not burn, and above 25 per cent they are fierce and easily started. Standing vegetation could never reach maturity. Evidence that oxygen was above 15 per cent two or three hundred million years ago comes from the discoveries of W G Challoner and his colleagues. They found, in the fossil record, layers of charcoal in sediments younger than this – charcoal that most probably came from fires in the vegetation, lit by lightning strikes. The presence of charcoal is therefore good evidence that oxygen has long been more than 15 per cent; the fossils of trees suggest that it can rarely have exceeded 25 per cent. It has probably been close to 21 per cent for several hundred million years. My colleagues often seem to find this sensitivity of the oxygen–fire relationship hard to believe. A modern illustration of its reality comes from experience with nuclear submarines.

These vessels stay under water for months on end and synthesize their atmospheres, making oxygen by the electrolysis of water. Submariners soon found that excursions of oxygen downward to even 15 per cent could be tolerated. An increase of oxygen of more than 1 per cent to 22 per cent, however, could not be allowed because of the greatly increased risk of fire it would bring.

So how is oxygen regulated? The answer depends upon the period. In the Archean there was almost none and regulation was not needed. In the Proterozoic, regulation was I think due to the slight but real toxicity of oxygen that set an upper limit of around 1 per cent abundance. For the last few hundred million years flammability has set the upper limit close to its present abundance of 21 per cent.

In more recent times, the existence of large animals and plants has made oxygen production easier and necessary and these factors have outweighed the disadvantages of oxygen toxicity. Oxygen toxicity is a major factor setting the lifespan of animals (see right), but adequate biological defence mechanisms have evolved to reduce the rate of damage by oxygen to acceptable levels.

Nitrogen

Nitrogen is the most abundant gas of the atmosphere, making up nearly 80 per cent of it. Nearly all the nitrogen of the planet is in the air. There is very little nitrogen anywhere else on Earth. The surface waters of the sea have barely enough for the organisms there, and on the land surface nitrogen is generally scarce except in living organisms. Yet, as I shall explain, one would expect to find most of Earth's nitrogen in the oceans – and only traces of it in the air. Is life responsible for this apparent anomaly?

In organisms, along with hydrogen, carbon and oxygen, nitrogen is a principal structural element. All protein and DNA molecules use nitrogen as a building block, as do many essential biochemicals such as vitamins and chemical messenger molecules. The construction of living organisms would be difficult without the peptide bond:

$$-C]-CO-NH-$$

At once strong, easily made, yet easy to disconnect, this is the same bond that chemists have used to make nylon and polyurethane: plastics that exploit the desirable mechanical properties of proteins.

In the Earth's environment, the most stable form of the element nitrogen is the nitrate ion dissolved in the ocean. (On Venus, which is a hot and dry planet, the stable form of nitrogen is the gas itself.)

In nitrogen gas the two atoms of the element are tied together by one of the strongest bonds known in chemistry:

$$N\equiv N$$

In the atmosphere it takes the high temperatures and fierce conditions within a lightning flash to break this bond and combine nitrogen with oxygen. (Nitrogen compounds make good explosives because the high energy used in their formation is released when two atoms of nitrogen recombine, and because they can flip with ease from a crystalline solid such as ammonium nitrate into a hot gas.)

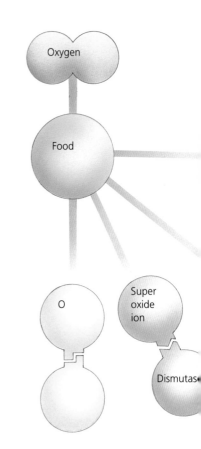

Detoxifying oxygen

Breathing oxygen is a hazardous business. Oxygen, like nuclear radiation, is a carcinogen and mutagen. The normal process of oxidative metabolism (the processing of food and water inside the cell in the presence of oxygen) produces a variety of "broken" chemicals called free radicals. These include the hydrogen and hydroxyl radicals (OH^-), the superoxide ion, and hydrogen peroxide (H_2O_2), all of which are capable of destroying the genetic instructions of the cell. In the long evolutionary period since the Archean, living systems have devised a range of counter-measures: antioxidants such as vitamin E (tocopherol) to scavenge the

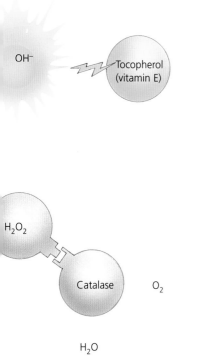

OH^-

Tocopherol
(vitamin E)

H_2O_2

Catalase O_2

H_2O

hydroxyl radicals; superoxide dismutase to destroy the superoxide ion; catalase to neutralize hydrogen peroxide, and others. Even so, breathing oxygen does set a limit to an organism's lifespan – small animals, such as mice, which have a much higher metabolic rate than us, have correspondingly shorter lifespans. The illustration above indicates some of the metabolic means of detoxifying oxygen that living cells have evolved. That oxygen should be a poison often comes as a surprise to the lay person accustomed to regarding oxygen as vital to life and health. The point in fact illustrates a truism of ecology and of medicine, "the poison is the dose".

Peptide, double, and other bonds

The peptide bond is important to living matter. It forms the links between the amino acids that make up a protein chain.

But what are bonds? Imagine an atom to be like a Lego brick, with projecting plugs and sockets that match and fit those of other bricks to form what chemists call covalent bonds. The bricks can also be electrically charged, in which case positive bricks are attracted to negative ones, to form ionic bonds. Atoms can also join by mixed bonds, which have some of the flavour of the plug-and-socket type and some electrical attraction. Nearly all matter on Earth is made up of atoms joined by any one of these three types of bond.

Atoms, like Lego bricks, are of different shapes and sizes. Hydrogen, for instance, has only one plug and one socket (it is monovalent), and is normally found as two atoms connected as a pair, H2, the hydrogen molecule. Oxygen, more complex, has two plugs and sockets (divalent), at right angles; when it combines with hydrogen to form water, H2O, the molecule is L-shaped. Nitrogen has three principal plugs and sockets (trivalent) arranged at angles, and carbon has four (tetravalent), pointing to the four corners of an imaginary regular pyramid – an odd shape for a building brick, but a most important shape in nature.

A pair of atoms can connect up with single, double, or triple bonds. The nitrogen gas molecule, N_2, for instance, has two atoms linked by a powerful triple bond. For double and triple bonds the extra bonds have to bend, because they do not point at each other directly.

Bending requires force. So some molecules with double or triple bonds have stored energy that makes them spring apart easily.

The peptide bond of the protein chain is

$$- C - N -$$
$$\quad \| \quad |$$
$$\quad O \quad H$$

The carbon atom has two of its plugs and sockets linked as a double bond with oxygen, and one linked singly to a nitrogen atom, which has a hydrogen atom attached. The free single bonds of the carbon and nitrogen link to the amino acids.

$$\text{amino acid} \begin{bmatrix} \end{bmatrix} \begin{array}{c} - C - N - \\ \| \quad | \\ O \quad H \end{array} \begin{bmatrix} \end{bmatrix} \text{amino acid}$$

Energy is often released when atoms come together to form molecules. Hydrogen and oxygen, for instance, burn explosively, to form water. A few substances have enough energy stored in their molecules to make them combustible even in the absence of oxygen; in some the stored energy is great enough to make the combustion explosive. Ammonium nitrate, for instance, decomposes explosively to form oxygen, nitrogen, and water:

$$NH_4 - NO_3 \rightarrow N_2 + O + 2H_2O$$

The formation of the triple bonded nitrogen molecule, N_2, or $N \equiv N$, is one of the most energetic reactions of chemistry. It is this energy, plus that released by water formation, that together make ammonium nitrate powerfully explosive.

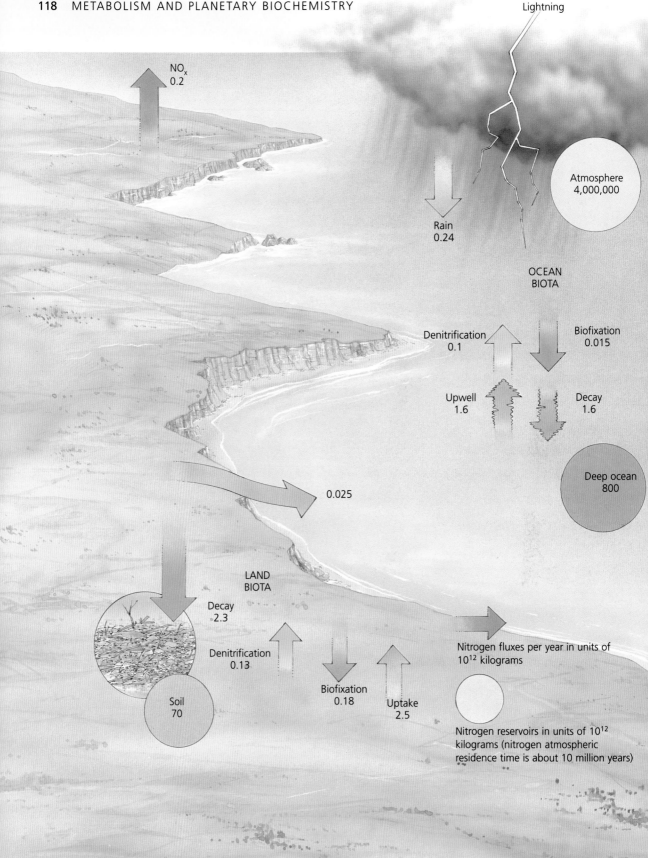

Lightning

NO$_x$
0.2

Atmosphere
4,000,000

Rain
0.24

OCEAN
BIOTA

Denitrification
0.1

Biofixation
0.015

Upwell
1.6

Decay
1.6

Deep ocean
800

0.025

LAND
BIOTA

Decay
2.3

Denitrification
0.13

Biofixation
0.18

Uptake
2.5

Soil
70

Nitrogen fluxes per year in units of
10^{12} kilograms

Nitrogen reservoirs in units of 10^{12}
kilograms (nitrogen atmospheric
residence time is about 10 million years)

The nitrogen cycle

Nitrogen is the one element found almost wholly in the atmosphere; in the natural world it is in short supply both on land and in the sea. If there were no life on Earth, but the oceans remained, all of the nitrogen would in a few million years move from the atmosphere to the oceans. Lightning flashes combine nitrogen with oxygen; the oxides react with water and hydroxyl radicals to form nitric acid, which falls in rain and is soon neutralized to form nitrates. In the absence of life, nitrates are quite stable and would lock up the element as nitrate ions dissolved in the oceans. But life reverses the flow. Nitrogen is essential to life, a key element in proteins and DNA. Biofixation of nitrogen (its capture and conversion to biological compounds by nitrogen-fixing microorganisms) ensures a constant supply of this nutritious element for uptake by land and sea biota. Other microorganisms, the de-nitrifying bacteria, work on the detritus of life and return nitrogen to the atmosphere – mostly as gaseous nitrogen but also as the gases nitric and nitrous oxide, and as ammonia. The total nitrogen flux is about 500 megatons a year.

If there were no life on Earth the continued action of lightning would eventually remove most of the nitrogen from the air and leave it as nitrate ions dissolved in the ocean. And when all the oxygen had gone, nitrogen would still react with carbon dioxide. In a lightning flash nitrogen also combines with carbon dioxide to form nitric oxide (NO), which eventually ends as nitric acid dissolved in raindrops. Carbon dioxide itself would be replenished from volcanoes and from the reaction of the nitric acid with carbonate rocks. There would also be some return of nitrogen gas from the sea by the reaction of the nitrate ion with the hot iron-rich rocks at the sea-floor spreading zones. On a lifeless Earth it seems probable that these inorganic forces would partition nitrogen so that most was in the sea and only a little was in the air.

Living organisms act continuously to pump nitrogen from the oceans and land surfaces back into the air. A fair proportion of the nitrogen returns to the air not as nitrogen gas but as the oxides of nitrogen, nitrous oxide ($N \equiv N_2 O$), and nitric oxide (NO). Scientists have not yet discovered what if any geophysiological roles these compounds have. I sometimes speculate on whether natural gaseous emissions such as those of nitrous oxide and methyl chloride serve to deplete stratospheric ozone. Too much ozone in the stratosphere could be as bad as, or worse than, too little.

From the viewpoint of planetary biochemistry, there are two good reasons for keeping nitrogen in the atmosphere. First, nitrogen sustains most of the atmospheric pressure and is a fine natural diluent. It is like the water added to the whisky to make it less fiery to drink. Second, the problem of oxygen and fires would become acute were nitrogen to be removed or even diminished. What determines flammability is the mixing ratio, the proportion of nitrogen and oxygen, not the pressure of oxygen. For animals it is the amount of partial pressure of oxygen that determines their survival. A fire can be lit on the top of Mount Everest as easily as at sea level, but no one can survive there without additional oxygen. I wonder also what would be the climatology of the Earth were nitrogen absent from the air. The greenhouse effect (see p. 137) would be less, not because nitrogen interacts with infrared like carbon dioxide, but because the greenhouse effect of carbon dioxide, and of other gases such as methane, is decreased as the total atmospheric pressure lessens.

The transfer of all the nitrogen from the air to the oceans would be very unwelcome to ocean organisms. Apart from the intrinsic toxicity of excess nitrate, the addition of the nitrogen as nitrate salts would increase the salinity of oceans by 33 per cent. This would raise ocean salinity above the critical limit for the cell membranes of mainstream life.

Trace gases in the air

The significance of an atmospheric gas does not depend on its abundance alone. Nitrogen is 78 per cent of the air and methane only 1.7 parts per million. At first glance methane might be regarded as just a trace gas, a chemical curiosity. Nothing could be further from the truth. Living organisms might manage without nitrogen, but life could not have developed, nor could it continue, without methane. As the case history at the end of this chapter shows (see p. 129), if methane production ceased, oxygen would rise to levels impossible for present life. Not only this, but nearly all the carbon would be taken out of circulation by burial.

A more realistic measure of the significance of a gas comes from its flux to and from the atmosphere. By this measure nitrogen and methane are equally significant, passing through at around half a gigaton (1 billion tons) a year.

The fluxes of the atmospheric gases are illustrated opposite. They fall into three groups. The greatest fluxes are of oxygen and carbon dioxide, at 100 and 140 gigatons a year respectively; then come nitrogen and methane at close to 0.5 gigatons a year; and lastly the gases nitrous oxide, dimethyl sulphide, methyl iodide, methyl chloride, carbon disulphide, and ozone at a lower flux of between 1 and 100 million tons a year. These proportions are not so different from the metabolic and biochemical fluxes through a human being. Oxygen and carbon dioxide dominate, then there is a lesser flux of nitrogen excreted as urea, and methane excreted in our farts. Some humans produce as much as 30 litres of methane a day, a curious fact that was discovered during the selection of candidate astronauts.

New discoveries

Intuitions coming from Gaia theory have led directly to the discovery of rare atmospheric gases with a potential role in planetary biochemistry. In 1976, for instance, I suspected, and confirmed, the presence of methyl chloride as a trace gas. It was the one-sided thinking of some campaigning environmentalists that spurred me to this discovery. In their view, all chlorine compounds were, like the notorious CFCs, to be classed with plutonium as products of industry or government agencies – and therefore bad. At a meeting in Washington on the depletion by CFCs of stratospheric ozone, it was proclaimed with great confidence that there could be no biological source of chlorine-bearing gases such as methyl chloride – indeed that there was no point in looking for methyl chloride, because "the biota never used chlorine in biosynthesis". I could not help wondering about the drugs chloramphenicol and chlortetracycline, both synthesized by microorganisms. I even wondered about common salt (sodium chloride). At the time I said nothing, but went home determined to find out for myself if methyl chloride was in the

Fluxes of atmospheric gases

The significance of a gas is more apparent from its flux – that is the rate at which it flows annually into and out of the atmosphere – than from its abundance, the level at which it is present. Nitrogen, for example, is by far the most abundant gas; but carbon dioxide has a much greater flux, with oxygen close behind. These are the two great metabolic gases of Gaia, and indeed the flux of a gas is some measure of its involvement in Gaia's self-regulatory system. The diagram (right) shows the principal metabolic and biochemical gases of Gaia, the relative fluxes of atmospheric gases expressed logarithmically compared with the parallel human metabolic and biochemical fluxes (below).

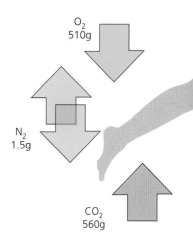

O_2
510g

N_2
1.5g

CO_2
560g

Human metabolic fluxes

That the chemistry of living organisms and planetary chemistry are part of a single domain is suggested by the parallels between the annual fluxes of Gaia's main gases (above right) and the daily chemical fluxes of the human body (above). In both, oxygen and carbon dioxide dominate, followed by methane and nitrogen.

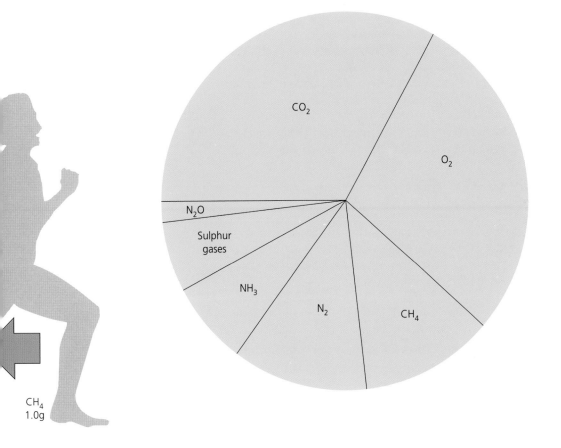

CH_4
1.0g

Flux versus abundance

The most convincing illustration of the significance of flux rather than abundance comes from considering the methyl radical. This gas, methane with one hydrogen missing, flows through the air at about 500 million tons a year – about the same flux as nitrogen. Yet it is so rare that there is in the air only one part in 10^{19}. The methyl radical is the product of the reaction of methane gas with the hydroxyl radical; it is the source of most of the hydrogen and carbon monoxide in the air.

The hydroxyl radical (.OH) is, fortunately, more abundant; yet even so, it is present at only one part in 10^{13} in the air.

Its annual flux, however, is even larger than that of methane. It is fiercely reactive. Were it available as a gas at atmospheric pressure it would set fire to and burn almost anything combustible. The hydroxyl radical is the great aerial scavenger and oxidizer. It removes methane from the air and sets its current low abundance. It converts the oxides of nitrogen to nitric acid, and sulphur gases to sulphuric acid. It cleanses the air of most pollutants. This remarkable reagent is also present in our bodies within white cells where it can be produced on demand to destroy invaders or scavenge debris.

air. A special new detector was required, which I made, and with it I found the gas to be present at nearly one part per billion, much more abundant than the CFCs in 1976. I reported this discovery in a *Nature* paper in 1977. The main sources and functions of methyl chloride are still unknown. Some comes from biosynthesis by fungi, some from the oceans, and some from forest fires and industry. It is an atmospheric gas that deserves more study.

Much more is known about three other trace gases, also discovered as a result of Gaia theory predictions. In the early 1970s I argued that there would be a need for biochemically synthesized gaseous compounds and sulphur and iodine. They would be needed to return these essential elements from the oceans, where they are abundant, to the land surfaces where they are scarce, yet necessary for life. (Iodine, for example, is an essential element for the thyroid hormones that regulate the rate of metabolism in many animal species.) The prediction was confirmed initially by the voyage of the RV *Shackleton* from the UK to Antarctica in 1972, during which I identified in sea-water samples the presence of the gases dimethyl sulphide (DMS), carbon disulphide, and methyl iodide. The confirmation of the presence of the two sulphur gases was particularly exciting, as it led to the perception of an extraordinary fact – the link between ocean algae and planetary cloud cover.

The sulphur story

For years, scientists had speculated about the sulphur cycle. Sulphur is regularly lost from the land, as sulphate ions in river run-off to the sea. The weathering of sulphur-bearing rocks, the sulphur extracted from the ground by plants, and the amounts put into the air by the burning of fossil fuels, are not nearly enough to compensate for the millions of tons washed into the sea each year. Without some mechanism for returning sulphur from the sea, land organisms would soon be starved of this essential element. So how is sulphur replenished?

Conventional scientific wisdom about the natural cycle of sulphur requires that large quantities of hydrogen sulphide be emitted from the oceans to make up for the losses of sulphur from the land. It was in the early 1970s that I began to question this explanation. As a one-time chemist I knew that hydrogen sulphide is rapidly oxidized in water containing oxygen, and also that it stinks – and so should be readily detectable – and therefore for both these reasons, I doubted if it were a major carrier of sulphur from the ocean to land. On the other hand, that elusive smell of the sea, for which the Japanese have a special word, is much like the sweet, ethereal, sulphury smell of dimethyl sulphide. Once you have smelt this gas, pleasant when diluted, it is recognizable ever after as a significant component of the aroma of fresh fish straight from the sea.

Earlier, while on holiday on the southwest coast of Ireland, I had had with me a paper by Professor Frederick Challenger, in which he observed that many marine organisms emitted DMS. Inspired by his findings, I had begun to conduct my own investigations. Equipped with a crude gas chromatograph, and a few simple pieces of scientific apparatus, I had found that nearly all algae emitted some of the DMS gas, but that *Polysiphonium fastigiata*, the red hairy alga that can be found festooned on the larger alga bladderwrack, particularly excelled in its emission.

By 1971, the idea of Gaia, the self-regulating planet, was well-established in my mind. I saw that the algal emissions of DMS were part of the planetary physiology and responsible for the transfer of sulphur from the ocean to land.

The findings of the *Shackleton* voyage that confirmed this proposal were published in *Nature*, but apart from the work of my friend Peter Liss who calculated from my data the fluxes of DMS and other gases to the atmosphere from the ocean, they were largely ignored. The German scientist, M O Andreae showed in the early 1980s that the output of DMS from the ocean was indeed sufficient to justify its role as the major carrier of sulphur from the sea to the land. Andreae not only confirmed my early work, but also showed that marine organisms emit vast quantities of DMS, particularly over the "desert" areas of the open oceans far away from the continental shelves. These findings led me and the meteorologists R Charlson and S Warren and Andi Andreae to make the exciting proposition that the rapid oxidation of DMS in the air over the ocean to form sulphuric acid droplets could provide the nuclei that are needed for the condensation of water vapours to form clouds (see p. 127).

Small droplets of sulphuric acid are ideal for this purpose, and over the oceans there is no other significant source of condensation nuclei from which to form clouds. (The aerosol of sea salt, which might be thought to nucleate cloud droplets, is much less efficient than the microdroplets of the sulphur acids.)

The oceans cover two thirds of the Earth's surface, and anything that affects cloud cover over these could significantly affect global climate. Whether this cloud–alga link plays a role in Gaia's self-regulating climate control mechanism is explored in the next chapter. But what we can ask here is: what selective advantage might the marine algae gain from the emission of DMS? They obviously do not return sulphur to the land solely for altruistic reasons, not produce the precursors of cloud condensation nuclei as part of some grand design to keep the planet cool. What could be the possible advantages?

A glimpse of how DMS production might have evolved is offered by the properties of a strange compound called dimethylsulphonio propionate. This substance is what organic chemists call a betaine (because they were first isolated from beets). Betaines are electrically neutral salts, and are important to the health of organisms living in a salty environment because they are non-toxic to cells in high concentration, unlike salt (sodium chloride) itself.

The production of dimethylsulphonio propionate may have been a cellular response to salt stress by marine algae drying out on some ancient shore. As water evaporated from their cells, the internal salt concentration would have risen above the lethal limit, and they would have died. In the way of evolution, those algae that had in their cells neutral solutes like the betaines would have been less desiccated and would have left more progeny. In time, the synthesis of betaines would have been common among all shore-living marine algae.

Successful inventions tend to spread. The concentration of salt in the sea is always uncomfortably high for living organisms. For the unicellular or small floating organisms in the open oceans, unable to regulate their internal salt by osmotic pressure, synthesizing sulphur betaines may have been the cheapest way, in terms of energy, of achieving a low salt interior. Dimethylsulphonio propionate would have been the natural choice, because sulphur is abundant in the ocean, whereas nitrogen (the basis of betaines in land plants) is not. This may not be the whole explanation of the presence of dimethylsulphonio propionate as the prominent algal betaine, but there is no doubt that the alga that contain it are the source of DMS.

The excretion of DMS to the atmosphere can bring the algae inadvertent benefits. The extra cloud cover from the presence of sulphuric acid nuclei (formed by DMS in the air) changes the local weather, increasing wind velocity and stirring the surface waters, mixing in the nutrient-rich layers beneath the depleted photosynthesizing zone. Continental dust particles when rained out on to the sea may also help the nutrition of the algae there. Finally, the clouds formed above the ocean filter the radiation reaching the water surface and reduce the proportion of potentially damaging short-wave ultraviolet light.

Taken together, these effects may be enough to improve the meadows of the sea and enable the algal species there to leave more progeny. The geophysiological system requires the continuing production of dimethylsulphonio propionate and DMS, and of the algae that make them.

Ocean biochemistry

Open any textbook on ocean science and the one thing not considered is water. Much will be written about the currents, the salinity, and the organisms of the oceans, but the great mother of a medium that makes it all possible is seldom mentioned. Water is so commonplace it has become something like life, taken for granted and never thought about.

Emiliana huxleyi x 10,000
This microscopic marine organism has a high output of DMS.

Marine algae and DMS

Marine algae are able to cope with salty conditions by the production of dimethylsulphonio propionate (DMSP). This substance is electrically neutral, carrying a positive charge associated with sulphur, and a negative charge associated with the propionic acid ion, on the same molecule. The internal neutralization of its ionic charges renders it non-toxic, and cells that are able to substitute a large proportion of betaine for salt are able to reduce the osmotic pressure between the cell interior and the external sea water.

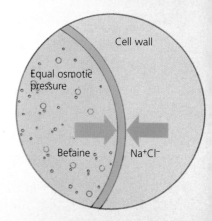

The sulphur cycle

When marine algae die or are eaten, the sulphur betaine decomposes easily to yield the acrylic acid ion and dimethyl sulphide (DMS). Onshore breezes carry the dimethyl sulphide inland where atmospheric gases decompose it into a non-sea salt sulphate NSS–SO_4^{2-} aerosol comprised of sulphate and methane sulphonate. In this form sulphur is deposited on the ground, thereby enhancing the growth of land plants, and also increasing the rate of rock weathering. The increased flow of nutrients to the oceans is of obvious benefit to marine organisms, and so the sulphur cycle is perceived as mutually beneficial to both land- and sea-based ecosystems. W D Hamilton and T Lenton have in a recent paper entitled *Spora and Gaia* suggested that the winds generated by DMS emission could assist algae to spread their spores to fresher ocean pastures. This could be the first step towards explaining how Darwinian natural selection and planetary self-regulation are linked.

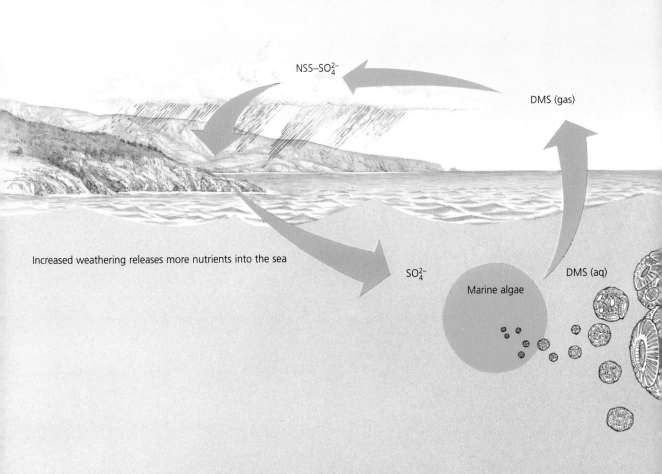

NSS–SO_4^{2-}

DMS (gas)

Increased weathering releases more nutrients into the sea

SO_4^{2-}

Marine algae

DMS (aq)

Climate feedback loop

The emission of DMS by marine algae to the atmosphere offers potential as part of a climate-control feedback loop. Rapid oxidation of DMS to form a non-sea salt sulphate aerosol provides the source of nuclei needed for the formation of clouds. The extra cloud cover increases the cloud albedo effect, causing a lowering in surface temperature, and increased surface wind levels. The lowered temperature may increase DMS output by the algae. The windiness is to their advantage, stirring surface waters to bring up more nutrients from deeper water.

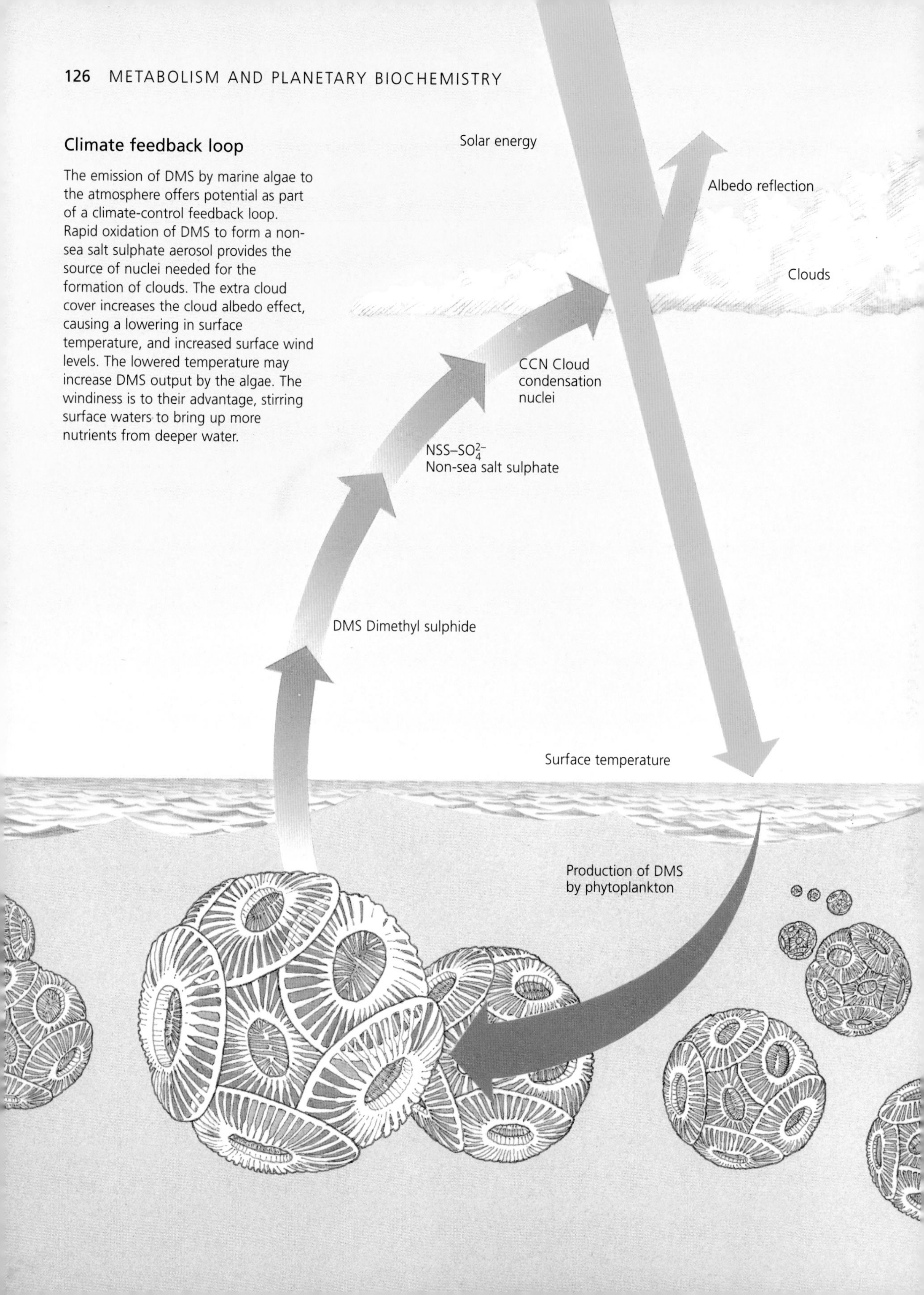

Solar energy

Albedo reflection

Clouds

CCN Cloud condensation nuclei

NSS–SO$_4^{2-}$ Non-sea salt sulphate

DMS Dimethyl sulphide

Surface temperature

Production of DMS by phytoplankton

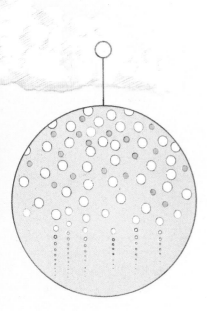

Cloud condensation nuclei

Clouds are formed when water vapour in the air condenses or freezes. But in order for the water vapour to condense, cloud condensing nuclei (CCN) must be present in the atmosphere. The major source of CCN over the oceans appears to be the dimethyl sulphide produced by planktonic algae in sea water.

In geophysiology, to ignore the water of the oceans would be nearly as foolish as ignoring the blood of an animal. Failing to recognize that sea water and blood, although fluid, not solid, are both tissues.

Without water there would be no life. We all acknowledge its beneficence and recognize that but for water our planet would be a barren rock ball like Mars or Venus. Like all things that bring comfort, we take water for granted.

It seems very probable that Earth, Mars, and Venus all had oceans soon after the turbulent Hadean period of the solar system. Why has the Earth alone kept its water? Venus is a planet very similar in mass to the Earth and it has outgassed almost the same quantity of carbon dioxide and nitrogen as has the Earth. It seems likely that it outgassed a similar quantity of water. For various reasons connected to its proximity to the Sun, Venus could not have enjoyed the leisurely evolution of the Earth. Venus underwent what my colleagues call, in an amazing mixed metaphor, a "runaway greenhouse". By this they mean an uncontrolled rise in temperature that ended with much of the ocean as steam. Water vapour is a potent greenhouse gas, and each increase in temperature increased the water vapour content of the Venus air. This in turn decreased the rate of heat loss to space and so led to an ever-rising surface temperature. The ultimate limit must have been when all of the ocean evaporated as steam.

At the high temperatures and pressures of the early Venus atmosphere, rock weathering may have been a rapid process, and may have taken the oxygen from water, thus leaving the hydrogen free to escape to space. In addition, water molecules near the edge of the atmosphere may have split into hydrogen and oxygen when illuminated by solar ultraviolet. This also would have provided a leak of hydrogen to space. The presence of abundant hydrogen rising up from the surface would have prevented oxygen from accumulating in the stratosphere to a level where its capacity to absorb ultraviolet radiation would have shielded the water molecules from dissociation. Venus would have lost its water probably in less than a billion years after its formation.

Mars also had water. On its arid surface it is now possible to see what appear to be dry river beds and what once were flood plains. These may be a fossil record of water on the planet over two billion years ago. No one knows how Mars lost its water, only that it seems to have done. The weathering processes that desiccated Venus would have operated on Mars at a slower rate, but Mars had much less water to start with. Perhaps Mars had life in the early stages and lost it in some catastrophe, such as a large planetesimal impact. Or perhaps life itself in the methane atmosphere of the Archean stage forced the rapid loss of hydrogen to space. (High levels of methane in the atmosphere increase the hydrogen content of the upper atmosphere and hence the rate of hydrogen loss to space.) Mars has only one third the gravity of Earth or Venus and hydrogen escapes much faster.

On the Earth these same processes would have been at work, and so why is our planet not now dry like Mars and Venus? The Earth has abundant oceans because it has evolved, not by geophysics and geochemistry alone, but as a system in which the organisms are an integral part. The evolution of this system has kept the planet moist. Water retention comes about by two processes (see also pp. 80–1). First, some hydrogen is fixed by microorganisms close to its source on the sea floor; they derive energy by combining it with sulphur to form sulphur compounds. (The hydrogen comes from the reaction of hot basaltic rock with high-pressure ocean water.) Second, the presence of atmospheric oxygen prevents hydrogen loss. Free oxygen is a source of hydroxyl radicals that rapidly scavenge hydrogen before it can reach the edge of space and escape. Oxygen also absorbs the ultraviolet radiation that otherwise would split water molecules. Together these processes so reduce hydrogen escape from the Earth that water loss was negligible after oxygen became the gas that dominated atmospheric chemistry.

The chloride ion

Keeping the Earth moist, keeping it suitable for organisms, is a cliff-hanging existence. If about 25 per cent of the water now in the oceans had dried away, ocean life would not be possible for most organisms. This is because the oceans would be over 6 per cent salt – far too saline for the structural integrity of the cell membranes (see p. 124).

The key element in determining ocean salinity is chlorine. Not the free element, a reactive choking poisonous gas, but the chloride ion. Ionized chlorine is a perfect spherical atom whose hunger for electrons is satisfied. Apart from its negative electrical charge, a chloride ion is unreactive like the atom of a noble gas. It is not absorbed by the particles of the ocean nor to any significant effect used by the organisms in their metabolism. Chloride ions just float around in the sea and by possessing a negative charge hold by electrostatic attraction a like number of positively charged sodium, potassium, and other ions.

There is a constant input of chloride ions to the sea, partly from rivers and partly from volcanic processes both above and below the ocean surface. The rate of this process is such that the salinity of the ocean would reach the critical 5 per cent salt level in about 800 million years. A long time but only about one quarter of the time that life has been on the planet. The persistence of ocean life suggests that the 5 per cent limit has not been exceeded. So where do the chloride ions go? Chloride ions leave the ocean trapped in the sediments and by exchange with other ions in the rocks of the sea floor. An important external sink for chloride ions is in evaporite deposits.

CASE HISTORY Inflammation

Could it be that a fiery battle for territory between oaks and conifers is Gaia's oxygen regulator?

Charcoal in the fossil record indicates that oxygen abundance has been close to present levels for hundreds of millions of years (see p. 113). Yet there have been periods during that time when oxygen *should* have varied. How could it have been regulated?

You might think forest fires provide an answer. Too much oxygen would mean more fires, hence fewer trees, hence less carbon to bury, hence less oxygen: fire as a negative feedback on oxygen. The flaw is charcoal: normally only 0.1 per cent of the carbon produced by plants is left for burial; but charcoal does not break down, and is all buried. Fire can convert up to 10 per cent of wood to charcoal, so that up to one hundred times *more* carbon would be buried if the fate of trees were burning: fires are a *positive* feedback on oxygen.

One intriguing clue comes from the fire ecology of eucalypts and conifers. These softwood trees are highly flammable. At first sight, their habit of depositing surplus branches, laden with resin, like prepared bonfires on the forest floor seems suicidal. Dry weather and a lightning flash would almost certainly start a fire. Such fires do start, but they tend to burn along the forest floor and do little damage to the standing trees. Conifers and eucalypts can survive fire; some conifer seeds even depend on fire for release from their wooden capsules. The fires however could serve to destroy competing species, such as oaks.

The fires set by the conifers and eucalypts burn completely, leaving little charcoal. And these softwoods also produce very little carbon for burial. But the oak is a hardwood, containing lignin, some of which survives burning and anaerobic fermentation and so more carbon is buried.

A possible oxygen regulatory mechanism emerges. If oaks become too dominant, more carbon is buried, oxygen increases, and so do the number and intensity of fires set by the conifers and eucalypts. These trees then succeed, less carbon is buried, and oxygen decreases. If the conifers and eucalypts are too successful, oxygen falls until the oaks are no longer curbed by forest fires, and so they begin to take over rival territory – and oxygen rises again.

The real process of oxygen regulation may be entirely different, or much more complex. (For instance, my friend the geochemist, Lee Kump, proposes a mechanism based on the spread of phosphorus, as an aerosol of ash from fires.) But I am sure forest fires are involved. No other phenomenon so tightly couples oxygen abundance to the growth and spread of organisms.

CASE HISTORY Methanogenemia

Science-fiction writers might care to consider a strange metabolic disorder of planets that could arise through an unfavourable mutation or by the meddling of an unwise molecular biologists. The disorder would be the failure of the methanogens to make methane from the organic detritus entering the sediments.

The disease could have come about because a well-intentioned scheme to convert waste organic matter into fuel and organic chemicals went badly wrong. Imagine an industrial molecular biologist who works to produce a specially tuned strain of methanogen that can convert detritus into products more valuable than methane. The biologist succeeds beyond all expectations – by engineering a strain that metabolizes vegetable waste, paper, and even some plastics into the perfect form of carbon, diamonds.

The new strain, *B. adamantanus*, although guarded with the greatest care, escapes down a laboratory sink. Before a week has passed, the vast anaerobic digester of the local sewage-treatment works is clogged with a grey crystalline powder that also contains the occasional large and perfect gemstone. The scene is set for the spread of the imaginary malady, methanogenemia.

If such on organism displaced the methanogens of the anoxic sector and organic carbon was turned wholly to diamonds, it could be a fatal condition for Gaia. The burial of all organic carbon would lead to a rise in oxygen of 1 per cent in about 30,000 years. Before long forest and plant fires would be out of control and even more carbon would be buried as charcoal. Land life would be curtailed but diamond burial would still go on in the ocean sediments. Oxygen would continue to rise until toxicity set a new equilibrium for oxygen at a level much higher than now. The loss of the land ecosystems would decrease the rate of carbon dioxide removal by weathering. The continued burial of some 2 per cent of carbon made by the plants, and the damage by fires and oxygen toxicity, would so limit the abundance of life that the system would be in danger of losing control of the planet. Whether the system survived or went into a marasmus-like decline would depend on the re-emergence of the methanogens.

Along the margins of the continents, lagoons often form by the relative movements of the land and the ocean. Sea water trapped in these lagoons evaporates in the Sun's heat and salt deposits form until eventually the whole lagoon is full of salt. Once formed, the salt of these lagoons becomes resistant to resolution. Soil and sediment eventually cover the surface of the lagoon and slowly a vast lump of sea salt is buried. The surplus salt of the oceans is scattered all over the Earth in the sedimentary rocks, sometimes, as in the Mediterranean, even beneath the sea. Geophysiologists ask the question, is the rate of salt removal regulated in some way by living organisms to keep the oceans safe for life? Or are the evaporite bed sinks formed entirely by chance?

Greg Hinkle and Lynn Margulis have made extensive studies of developing evaporite beds, as, for example, those found in the shallow lagoons of Baja, California. These lagoons are covered with bacterial mats, complex communities of microbes little different from those that must have flourished on Earth from the Archean age onward. The microbial mat communities form layers, with green and red microbes at the surface, anaerobic bacteria and the fibrous remains of earlier communities beneath, and the evaporated beds of deposited salts below. Hinkle and Margulis found evidence that the presence of bacteria at the lagoon surface did load the dice in favour of salt retention in the lagoon by, for example, coating the surface of the salt crystals with a water-repellent varnish coat that made them resistant to re-solution in rain water.

I have often wondered if something more was needed to bury salt than this purely local activity. One of the most intriguing speculations to arise from Gaia is that of a link between plate tectonics and the presence of life on Earth. The eminent geophysicist, Don Anderson, has stated that the formation of limestone deposits by organisms may have been important in triggering the change of composition of lithospheric rocks – a change that allowed plate formation and motion (see illustration). Could it be that the formation of lagoons destined to become evaporite beds is part of a Gaian plate tectonic mechanism?

The rain of calcium carbonate shells from ocean algae, for instance, occurs most abundantly in shallow water around the edges of continents. And it is here, too, that giant colonies of bacteria formed the limestone structures, the stromatolites mentioned in Chapter 5. As they grew, these stromatolites would have gradually sealed off portions of the shore to form lagoons, which eventually began to dry out as evaporite beds, with their microbial mat communities. The huge mass of these limestone reefs could also have contributed to the process Don Anderson describes – whereby the weight of accumulated limestone began to affect the chemistry of Earth's plastic crust beneath, so that plate movements and rock

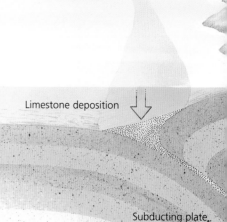

Ocean surface

Limestone deposition

Subducting plate

Plate tectonics and Gaia

Could Gaia be, even partially, responsible for the plate tectonic movements of the Earth's crust? The Earth is certainly, it seems, exceptional among the known planets in having such movements. It was the geologist Don Anderson who first speculated that the sedimentation of limestone on the ocean floor, long ago in life's history, could have so altered the chemistry and temperature of the crustal rocks as to make it possible for the machinery of plate movement to begin. The geological event that changed the crustal rocks near the continental margins and made the crust unstable is known as the "basalt–eclogite phase change". The early presence of the microflora of the oceans, and their constant rain of limestone tests to form limestone deposits, thus may have been the trigger that led to the slow swirl of the crustal plates across the globe.

folding took place around the continental margins, allowing the formation of more lagoons.

These salt-segregation processes seem always to have balanced the return of salt to the ocean by erosion and run-off, so that ocean salinity has remained fit for life.

The land surfaces

The land surfaces now cover 33 per cent of the Earth's area but house about half of the biota. Most of the land surface was once below the sea; the remainder is volcanic in origin. The thick layers of ocean sediments are all either wholly biological in origin (such as the vast deposits of coccolithophore shells that became chalk) or have been modified by the organisms (such as sandstones and clays). These ocean sediments are continuously being lifted up on to the land, there to be weathered and washed back into the ocean by rain. The fall of rain is so great that it is enough to fill the oceans in only 60,000 years. The rate of erosion of the land by the rain is so rapid that sufficient solid material to fill the ocean basins would be brought down in 17 million years. Although 17 million years may seem a long time, it is only three times longer than humans have been on Earth, and only one three-hundredth the age of the Earth. The ocean basins do not fill up because uplift removes the accumulating materials, or they are dragged below the edges of the continents by the movement of the plates. Moving plates cover the ocean floor and float, like ice floes in the Arctic, over a planet-wide ocean of magma. As these plates encounter the more solid islands of the continents, they slide down beneath, dragging sediment with them.

Thus there is a slow but continuous exchange of materials between the hot interior, the magma, and the surface environments of the Earth. The sediments that are dragged underneath the continent by plate movements are either a direct product of life – such as the chalky and silica tests of alga – or are largely modulated by life. These sediments are mixed in with the subductal crust and the magma. Fresh volcanic outpourings, the upwelling of magma under the oceans at plate junctions, and the uplift of buried rocks by tectonic activity, all replenish the surface with fresh materials.

Life on the land surfaces has always faced different and more severe problems than those of life in the sea. Desiccation is an obvious threat, for the fall of rain is fickle. The Sun's radiation both in the visible and the ultraviolet is more intense on the land. Here the temperatures encountered may be as high as 60°C or as low as –50°C. In the gentler environment of the ocean, the temperature range is from about 30°C to –1°C. In the seas, life can spread in three dimensions, but on land, spread away from the surface itself is restricted. In addition to these problems, elements essential to life such as iodine, sulphur, molybdenum, and magnesium are scarce locally on the land. There are also fires, droughts, cold spells, floods, and, in recent times, people.

Life has adapted to these more serious threats to existence on the land. The system of life-and-its-environment has profoundly adapted the land to favour the continuation of life. Consider, for example, how eco-systems are involved in the self-regulation of local climate – rainforests by evapotranspiring water and increasing local rainfall; conifer forests by absorbing spring and autumn sunlight and by shedding snow, so warming their near arctic regions.

Perhaps the most important of these adaptations of the environment was the conversion of regolith, the rubble and dust of a dead planet, to soil. Soil adds an extra dimension to surface life and allows the coexistence of anaerobic Archean-like life as a substratum to air-breathing life above. The soil has an atmosphere all its own with carbon dioxide 30 times more abundant than in the free air, and gases such as methane and nitrous oxide greatly enriched.

Soil plays a fundamental role in Gaia's planetary metabolism. The high carbon dioxide level is crucial to the weathering of calcium silicate rock, which, as said earlier, is the only true sink for carbon dioxide – the sink that prevents this gas from building up to levels uncomfortable for life. Carbon is buried in the soil and in the sediments of the ocean floor, and this burial allows the presence of free atmospheric oxygen. The anaerobic sediments of the soil are also the source of the methane that removes oxygen from the air, so keeping the balance, and is essential to life on Earth.

Atmosphere, oceans and land – in each we have seen how Gaia, the system of life and its environment, is involved in metabolic processes that link the biochemistry of organisms with the planet-wide biochemistry of Earth. The next chapter examines how these processes actively regulate Earth's climate – and have long been involved in maintaining conditions suitable for life.

CHAPTER SEVEN

Physiology and climate regulation

One of the great puzzles about the Earth is the constancy and comfort of the climate for whatever was the contemporary biota. Life began at least 3.8 billion years ago, and it must have been warm enough then. Otherwise we should not be here to talk about it now. The microorganisms living today at the surface and fixing sunlight to make food for the consumers are very little different in form from microfossils in rocks of that Archean period. And the organisms now living in the muds and sediments at the bottom of the lakes, marshes, rivers, and the oceans also bear a close resemblance to the microfossils of sediments of that same ancient time. The prevalence of these organisms throughout the geological record, and the abundant evidence from the rocks of the presence of liquid water, all tell that the Earth has neither frozen nor boiled nor been too hot for familiar forms of life.

The puzzle is that over the same period the Sun, in the natural course of its ageing, has increased its output of heat by about 25 per cent. If the Sun were to cool so that its output were 25 per cent less than now, then the mean temperature of the Earth would not be 14°C like now but would fall to somewhere below freezing. By the same argument, if the Earth 3.8 billion years ago was as warm as it is now, then a 25 per cent increase in solar output would have raised the mean temperature above 30°C, far too hot for comfort. We know that the Earth was not frozen in the Archean period and we know that it is not overheated now. To add to the difficulty D Schwartzman recently proposed that it was warmer in the first two eons of Gaia's existence. So what is the explanation?

The first attempt to answer the puzzle came from the astronomer, Carl Sagan. He and his colleague G Mullen suggested that the early Earth was warm enough for life on account of a greenhouse blanket of ammonia gas. This notion was hard to sustain because ammonia is rapidly destroyed by solar ultraviolet radiation and very large sources of ammonia would have been needed to maintain a sufficient gaseous greenhouse. But the idea stimulated other scientists, notably M H Hart and T Owen, to propose that the carbon dioxide of the early Earth atmosphere was much higher than now, and that this provided the warmth needed, through the greenhouse effect.

So what is the scientific basis of this greenhouse effect? And how does it work? It was first commented on by the British scientist John Tyndall in the 1860s, and discovered in depth by two other scientists, the Swede, Svante Arrhenius, and the American geologist, Thomas Chamberlain, some 30 years later. They wondered why the Earth is as warm as it is. If you calculate how much heat the Earth receives and how much it is radiating away as long wavelength infrared to space, you can work out the mean surface temperature. If the Earth today had no gaseous greenhouse to retain heat, the surface temperature would be −19°C, 33 degrees cooler than the present average comfortable temperature of 14°C. It occurred to these scientists that some of the gases in the atmosphere, in particular carbon dioxide and water vapour, when present at the thickness of the entire atmosphere, could absorb long wavelength infrared, and reflect it back to Earth, in a way like but not the same as the panes of glass in a greenhouse (see illustration opposite). This effect, as we now know, is one of the principal factors of the Earth's climate.

Hart and Owen's proposition of an early Earth with atmospheric carbon dioxide was fine for a start. Something like 10 per cent carbon dioxide in the air would have kept the Earth warm in spite of a cool Sun. But it only postponed the puzzle. If the early CO_2 level had persisted, we would have been very uncomfortable indeed now. But we know that 10 per cent carbon dioxide did not persist, as it is now 300 times less than that. How did this occur?

The abundance of carbon dioxide in the air depends on the balance between the amount being injected from beneath the crust through volcanoes and the amount lost from the air by chemical reactions at the Earth's surface. The volcanic source was three times larger at the start of life, because the young Earth was more radioactive than it is now, and there was more internal heating from the energy released by the decaying atoms. The decline of the input from the beginning until now is likely to have been a regular progression, punctuated only by violent events like planetesimal impacts or periods of intense volcanism. But this three-fold reduction cannot account for the 300-fold decrease in atmospheric CO_2.

Jim Walker and Jim Kasting, both geochemists, then proposed that as the Sun warmed up, carbon dioxide fell in abundance as a result of increased removal by rock weathering and by increased rainfall. They maintained that there was (and still is) only one sink for carbon dioxide in the air and that was the reaction (discussed in Chapter 6) of the gas with the rocks to form limestone, calcium carbonate. The weathering of the rocks by rain containing dissolved carbon dioxide was the practical expression of this sink. They assumed that living organisms had little or no effect on carbon

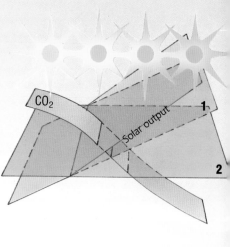

1 Earth temperature without life
2 Earth temperature with life

Solar output and temperature

Our Sun is an average star, and like most other stars increases its output of energy, or luminosity, as it grows older. This behaviour is due to a kind of internal greenhouse effect. Stars generate heat by fusing hydrogen atoms in their hot centres, to make helium. This gas behaves in a star rather like carbon dioxide in our planet's atmosphere: it acts as a blanket to impede the escape of heat, so raising the temperature of the star's interior. This greatly increases the star's fusion reaction and so its heat output. In the 4.6 billion years of our Sun's existence, it will have increased its heat production by 25 per cent, as the graph above shows. The dotted line (1) shows the effect this should have had on Earth's surface temperature. The solid lower plane (2) shows the actual history of Earth's surface temperature, so far as it is known. The difference between the two is considerable, and is an important part of the evidence for Gaia. It suggests that some temperature regulating effect due to life's involvement is at work – probably the level of the biologically controlled greenhouse gas, carbon dioxide. As shown on the graph, this has steadily fallen over the period of life on Earth, compensating for the increased solar heat reaching the planet.

Sunlight

Infrared

How does the greenhouse work?

Have you ever felt that sense of surprise when entering a greenhouse on a frosty but sunny day? The air inside is quite warm, even though there is no heating, warm enough for plants to grow. The Earth is a bit like a greenhouse. Space outside is very cold, yet at the Earth's surface it is everywhere, except in the wintry regions, warm enough for plants to grow. So how does it work?

Inside an ordinary greenhouse, with panes of glass, it is warmer than outside because all of the heat energy of sunlight, or nearly all, is in the visible part of the spectrum. Surprising as it may seem, light is heat. The light, of course, goes through the glass easily. Inside, it is absorbed by the green plants, by the soil, and by anything dark that is there. The absorption of the sunlight by the plants warms them and they in turn warm the air as it caresses them in its restless motion. They also lose heat by radiation: not by radiating light, because they are not hot enough, but by radiating at long wavelengths of what is called the infrared – wavelengths longer than those emitted by a red-hot object. (The wavelength of visible light is just a little less than 0.001 mm and that of heat from plants in a greenhouse about 0.01 mm.) The glass of the greenhouse is warmed by the air and by the long wave radiation from the plants.

Now the greenhouse as it warms begins to lose heat, both by warming the frosty air that passes over its surface (a process called convection), and by radiating heat at long wavelengths in the infrared. Eventually a balance is struck between the heat received and the heat lost, which leaves the air inside the greenhouse considerably warmer than the air outside. If the greenhouse were built of infrared transparent glass it would still be warm inside, but if the glass were not there at all, the warm air parcel inside would be blown away. So the greenhouse stays warm by retaining the heat it receives within it. As you know, leaving the door open frustrates this function and the greenhouse cools.

So what is the connection between a greenhouse and the greenhouse gases? There aren't any panes of glass in the sky – the sky is just transparent air going all the way out into space, getting thinner and thinner as it goes. The atmosphere is not solid and can mix, so it cannot act like the greenhouse to keep the parcel of air within from mixing with the cold outside air.

What happens is that Earth's greenhouse gases such as CO_2, water vapour, and methane in the air from the surface, where it is thick, absorb outgoing infrared, and the air warms up. Some of the warmth is re-radiated back downward and warms the surface of the ground, so less escapes to space.

It is this process of absorption of heat by the air itself in long wave infrared, as the warmth of the Earth goes upward, and its reflection back downward, that is called the greenhouse effect.

continues overleaf

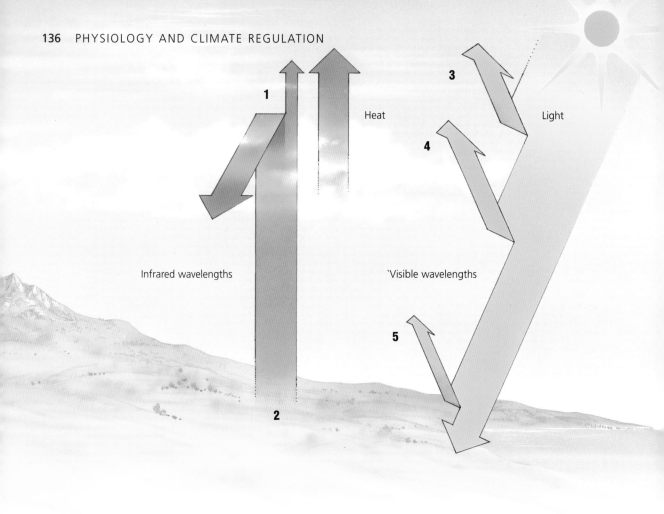

1

Heat

3

Light

4

Infrared wavelengths

`Visible wavelengths

5

2

It is not the same as, but it is very like, the greenhouse in your garden.

An unexpected consequence of the gaseous greenhouse is that the air high up in the stratosphere is cooler, in fact a great deal cooler. For a 2°C rise in temperature due to the greenhouse at the surface there may be a 20°C fall in stratospheric temperature. Just as the lagging around an insulated steam pipe is cool compared with the metal surface of the pipe, so the Earth with a greenhouse lagging has a cooler outer atmospheric surface than it would have unlagged.

The greenhouse gases

Greenhouse gases, such as carbon dioxide, are different from the main gases that make up 99.97 per cent of the air – nitrogen, oxygen, and argon. The first two of these are diatomic (made of two atoms joined together); argon is monatomic – just a bare atom. These gases cannot absorb either the infrared or the visible wavelengths of the Sun. Consequently, they neither warm the air as sunlight passes through it nor heat up as the infrared radiates back up.

But gases that have three or more atoms, such as carbon dioxide and water vapour, can absorb infrared radiation because their larger molecules naturally vibrate, at frequencies the same as the infrared.

The three-atom molecule of water, for instance, is joined so that one oxygen atom links two hydrogen atoms at an angle of 105 degrees, like a boomerang. This molecule can rotate or can bend to a sharper or wider V. The frequency of the motions coincides with the frequency of the infrared from the Earth. Thus water molecules resonate with the infrared and absorb its energy, and this keeps the air warm. A molecule of water vapour is a bit like a television antenna. The TV antenna receives signals sent by the transmitter at the wavelength that it's cut to and can also transmit at this frequency. Different antennae operate at different frequencies: a small 15–30cm antenna on a roof is

The greenhouse effect

The diagram, left, shows how the Sun's radiant energy reaches the Earth, mainly in the visible wavelengths of the spectrum, as light. Much is reflected back to space by the atmosphere (3) but some passes through to the surface. Lighter areas reflect this light back (5), but darker ones absorb it and radiate up energy in the longer infrared wavelengths (2). The greenhouse gases, CO_2, water, and methane, resonate with infrared, absorbing heat and re-radiating it, both up and down (1). In addition, low clouds reflect sunlight back to space (cooling) (4), while high ones also reflect heat back to Earth (warming) (1). The balance between heat lost to space and heat retained leaves a warmer atmosphere.

probably connected to a TV set; but a large array probably belongs to a radio ham operating on the much longer radio wavelengths. Similarly, different molecules, such as water vapour or carbon dioxide, resonate at different infrared wavelengths. Another gas that absorbs infrared wavelengths in the air, and is very important, is methane or natural gas. The methane molecule has one carbon atom with four hydrogen atoms connected to it at the corners of a tetrahedron, in structure like a pyramid. It can rotate and the hydrogens can stretch in and out, nearer to or further from the carbon, motions that resonate with infrared.

dioxide levels since, as we have seen, what was made by the consumers was taken back by the plants, and that the producers and consumers were always in balance. This was a nice idea but it still failed to account adequately for the great decline in carbon dioxide over geological time. Something more was needed.

In the early 1980s I proposed that Walker and Kasting's model could be brought alive by including the organisms in it. The idea of a pre-life atmosphere rich in carbon dioxide, and of the level of this gas as a key element in climate history, is very consistent with Gaia theory. I see carbon dioxide as a gas that is actively regulated. As the previous chapter showed, the intervention of living organisms in the great cycle of carbon is intricate and of global scale.

All life forms on the land surface, from microorganisms to trees, from amoebae to elephants, in various ways increase the rate of rock weathering, which is the sink for the greenhouse gas, carbon dioxide. And the ocean algae also pump down carbon dioxide from the air. First, in their rapid springtime growth, they use up the dissolved carbon dioxide at the ocean surface. Carbon dioxide then flows in from the air by diffusion and so some of the atmospheric burden of carbon dioxide is removed. Later, blooms of algae, the coccolithophores, use the carbon dioxide in the ocean to make their calcium carbonate shells, which then fall to the ocean depths. This is not by itself a sink for carbon dioxide but is part of the large life-driven process that removes carbon dioxide from the air and deposits it as calcium carbonate on the ocean floor. (See Chapter 6, pp. 108–9.) Little though we may yet understand the long constancy of the Earth's climate, life is so intricately bound up in the carbon cycle that it must be included in any theory.

Making models of the Earth's early climate

The present cool climate in spite of a warm Sun and the constant decline of carbon dioxide over the ages cannot be explained by either biology, geophysics, or geochemistry taken separately. However, a geophysiological model of the Earth with the evolution of the organisms and the rocks and air taken as a tightly coupled system does offer a plausible explanation.

To try to gain insight into the regulation of the atmosphere and the climate of the early Earth, I made such a geophysiological model. This model was a direct descendant of Daisyworld but considerably more complicated. My model planet was still circling its star and the model started back in early times when the star was cool. The input of carbon dioxide from volcanoes was taken to be three times larger than now, and the solar input some 25 per cent less than now.

Instead of daisies, the planet had an ecosystem made up princi-pally of primitive plants (that is, photosynthesizing bacteria). It was assumed that these plants took in carbon dioxide and turned it into the carbonaceous matter of their bodies, giving off oxygen as they did so; but as Chapter 6 showed, oxygen would not have accumulated in the air because the early Earth was full of reducing substances. There would therefore have been little opportunity for consumers to flourish (see pp. 111–13) A few consumers were nonetheless included in the model because it seemed likely that close to the plants there may have been pockets of oxygen and this would have provided local niches for them to evolve. The organic matter of the plants was principally turned over by the fermenters (methanogens) who converted the carbonaceous material of the plant debris into carbon dioxide and methane. In the model, this simple ecosystem of plants, fermenters, and consumers, was coupled into the evolution of the chemistry and physics of the planet. The growth of all the organisms was assumed to be limited by temperature in much the same way as the growth of the daisies was.

The climate of the planet was determined by the proportion of greenhouse gases, carbon dioxide and methane, in the atmosphere. The photosynthesizers acted like white daisies to cool the planet by pumping carbon dioxide from the air, the methanogens like dark daisies to warm the planet by adding methane to the air. Unlike Daisyworld where the dark- and light-coloured daisies compete in this model the abundance of methanogens was directly dependent on the abundance of the photosynthesizers. In consequence it behaves like a planet populated by a single dark daisy and warms as the sun warms. Like Daisyworld, the climate of my model planet rapidly became comfortable soon after the infant bacterial life settled down, and the gases of the air reached a realistic and comfortable steady state. The model predicted a methane abundance of between 0.01 and 0.1 per cent, and a carbon dioxide abundance of 0.5 to 1 per cent. (The steep fall of carbon dioxide abundance, from around 10 per cent at the start to 0.5 to 1.0 per cent at the steady state, was due to the greatly increased rate of rock weathering in the presence of organisms. The rest of the atmosphere was assumed to be nitrogen, argon and water vapour.) This result was consistent with the idea that methane could have been the chemically dominant gas after life commenced on Earth, in an atmosphere where oxygen was present only as a trace gas. That is the reverse of the situation now when oxygen is dominant and there are only traces of methane.

An atmosphere dominated by methane after life began is a prediction from Gaia theory and one that neither geochemists nor biologists working alone would have made. It provided a good test for the theory (see p. 26, the Table of Predictions).

Methane is a more potent greenhouse gas than carbon dioxide. The levels of atmospheric methane proposed would have been ample to have kept the Earth warm in spite of a solar output of 25 per cent less than now.

The basis of the Archean model

Like Daisyworld, my model of the evolution of Earth's atmosphere was a program – a set of mathematical formulae and instructions – run on a computer, and based on certain assumptions. The geochemistry of the Archean period was based on H D Holland's book, *The Chemical Evolution of the Atmosphere and the Oceans*. Solar input at the start was taken to be 25 per cent less than now, and with carbon dioxide input from volcanoes three times higher than today.

Instead of dark and light daisies, rabbits, and foxes, the model had the three primitive ecosystems of the Archean – photosynthesizing bacteria drawing on the abundant carbon dioxide to manufacture organic matter; a few consumers in the pockets of free oxygen they generated; and fermenters (methanogens) using the decaying organic material from both.

The bulk of the atmosphere was assumed to be nitrogen, with carbon dioxide initially around 10 per cent, giving a temperature of around 28°C. The Sun was allowed to warm up as it aged, and climate regulation was based on the greenhouse effects of methane and carbon dioxide, with the weathering sink for carbon dioxide increasing with the growth of organisms.

Organism growth was assumed to be slow below 5°C and above 50°C, and optimum near 25°C. It would also be dependent upon the abundance of carbon dioxide. Oxygen would have a positive effect on growth at low levels, through its ability to increase the rate of release of nutrient elements from the rocks, but it would have a poisonous effect at high levels.

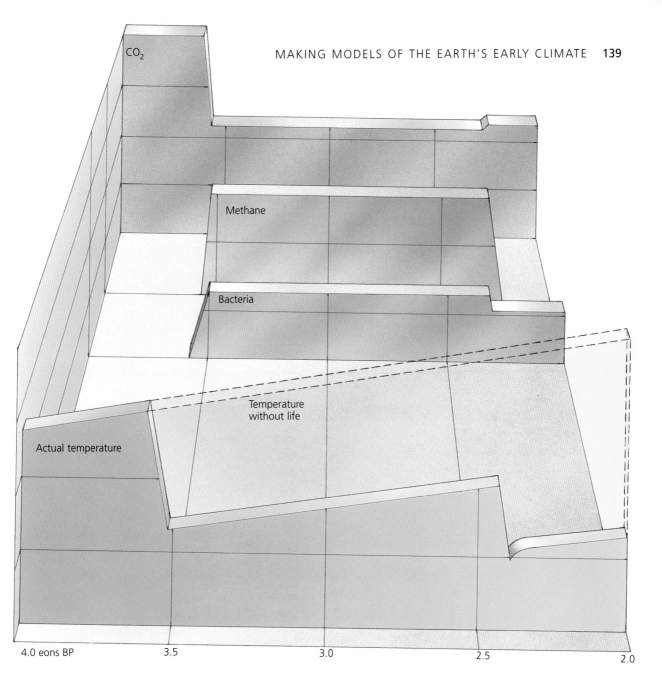

CO$_2$

Methane

Bacteria

Temperature without life

Actual temperature

| 4.0 eons BP | 3.5 | 3.0 | 2.5 | 2.0 |

The results of the Archean model

The model results match what is known of the Earth's early history.

The climate graph (lower) shows how, without life, the warming Sun would have led to a steadily warming Earth (dotted line). The solid line on the same graph shows how life changed all this. There is an abrupt fall of temperature after life starts, due to a rapid decline of carbon dioxide as it was used up by the photosynthesizers. The temperature then stabilizes, rising slowly throughout the Archean. The upper graphs of the gases and the bacterial populations show why – methane from the fermenters was accumulating in the atmosphere and its greenhouse effect replacing that of carbon dioxide. The temperature suddenly falls once more at the end of the Archean, when the sudden appearance of free oxygen marks the decline of methane.

A rapid reduction of carbon dioxide abundance after life began is consistent with the Earth's rock weathering record. The Earth's temperature was fairly stable in the Archean but a cold glacial interlude about 2.3 eons ago may have marked its end. The model matches this pattern.

Like Daisyworld, this model shows an abrupt change of conditions as soon as life starts. The organisms grow and change their environment and the atmosphere rapidly, until a steady state is reached and Gaia runs on in comfortable homeostasis.

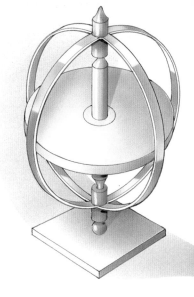

In the model, as in the Earth's history, the Archean did not persist indefinitely. The calm was disturbed by the sudden emergence of oxygen as the chemically dominant atmospheric gas. When the oxygen abundance approached twice that of methane, the atmosphere became oxidizing, as it is now. (Only when oxygen became dominant could the ozone layer that shields the Earth's surface from much ultraviolet radiation from the Sun take form; there may, however, have been an Archean equivalent, see below, right.)

This simple model of the Archean and the transition to the Proterozoic is stable and resistant to perturbation both internal and external. It suggests plausible levels for climate and for three atmospheric gases (oxygen, carbon dioxide, and methane), and includes three coexisting ecosystems (photosynthesizers, methanogens, and consumers).

There is no way to build such models from conventional biogeochemistry. The geosphere and the biosphere should not be studied as separate entities except for the classification of data. The operating system of the Earth behaves like one superorganism, Gaia. An organism that, through homeostatic feedback between life and its environment, has long maintained a climate comfortable for life.

Homeostatic responses

One good test of a homeostatic system is its ability to respond to disturbance and injury. Gaia has survived at least 3.8 billion years, and in this long life she has experienced many challenges, including the upset of a complete change of atmospheric state, when oxygen became dominant. She has also suffered numerous diseases, some of which I have described in Case Histories in earlier chapters. But of all the stresses on Gaia's homeostatic functions, none compares in severity with that of injury by planetesimal impacts.

The collision of one of these monster missiles makes the Earth shake like a jelly and leaves a crater up to 300 kilometres in diameter. No other events in Gaia's history have been so injurious, and there have been many such blows causing shock and the threat of marasmus (system failure, see p. 69), and scars that take millions of years to heal. The continuation of life and the constancy of the climate in spite of them is a tribute to the resilience of the system, and is also among the more convincing items of evidence to suggest the presence of a self-healing and self-regulating homeostatic organism called Gaia.

You may think I exaggerate about the frequency and destructive power of these planetary missiles. If you do, look at the map of Canada, overleaf. The map itself shows that the hard rocks of Canada, were it not for the continuous smoothing action of the weather, would look just like the surface of the Moon or Mars.

The large impact most discussed is the most recent, one that occurred 65 million years ago. The consequences of this disaster

The Archean "ozone layer"

In a methane-dominated atmosphere there could not have been an ozone layer. You might think that this would have exposed the Earth's surface to a flood of lethal ultraviolet radiation, and life of any form would have been impossible. The evidence suggests otherwise. The tough bacterial life of the Archean is unlikely to have been bothered by ultraviolet. Even were it sensitive, a methane-dominated atmosphere would have included an organic chemical equivalent of ozone. Methane is chemically very stable and in an atmosphere free of oxygen would have decomposed only in the upper atmosphere where short wave ultraviolet penetrates. Here, in a complex sequence of reactions, a small but significant proportion of methane would have gone to form organic polymers that could absorb ultraviolet, just as ozone does now.

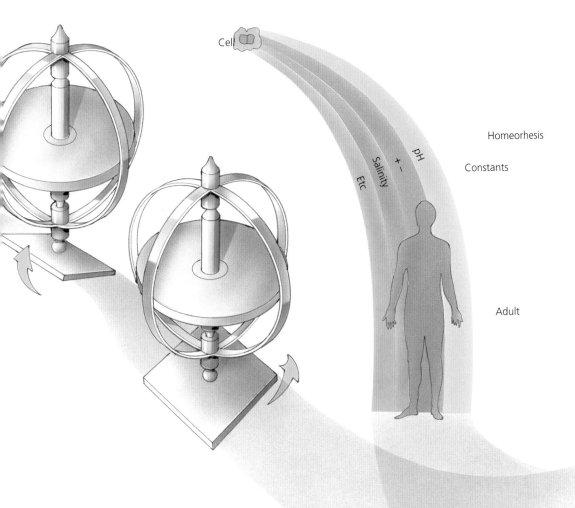

Homeostasis and homeorhesis

The fact that the amount of the gases oxygen, carbon dioxide, and methane, varies with time is often used as a criticism of Gaia theory. Gaia, the critics say, is claimed to keep constancy, to be homeostatic, yet the three most important atmospheric gases have varied widely in abundance during the history of the Earth. This common criticism is in error and arises from a failure to understand what is meant by the physiological term homeostasis. Homeostasis in living systems is not a permanent, fixed state of constancy; it is a dynamic state of constancy.

A spinning gyroscope, for instance, is resistant to perturbation only so long as its speed of rotation is sufficient. If it spins too slowly, it will wobble and fail. But a living homeostatic system, while it too can "wobble", can avoid failure by moving to a new state of constancy, and reset itself to new limits. A closer analogy to this is the operation of a ship's autopilot device. Given a course, it will steer the ship faithfully despite any perturbation by wind or currents. But if major storms or rocks lie ahead, the ship may need to deviate to a different course. When this is set, the autopilot rapidly changes the ship's direction of motion to a new, stable motion,

and homeostasis then proceeds. This process of sudden moves from one stable state to a new stable state is known as *homeorrhesis*.

It was the geneticist C H Waddington who coined this term, to describe the common property of living things that change while staying constant. He was thinking of the controlled growth of a fertilized egg, which proceeds by stages to an adult organism. The mass, form, and functions change; but pH, ionic strength and many other properties remain constant throughout (see diagram above).

Gaia's history is similarly characterized by homeorrhesis with periods of constancy punctuated by shifts to new, different states of constancy. With some variables, such as temperature, the changes are small – such as the 5°C difference between a glaciation and an interglacial. With others, such as gaseous abundance, the levels of homeostasis have progressively changed in a series of steps. Carbon dioxide has tended downward at each step, oxygen upward; methane has risen and then fallen. Yet the system has maintained conditions comfortable for life.

have been the subject of a heated and entertaining scientific debate concerning the demise of the dinosaurs (see p. 144). This particular planetesimal blow is also very revealing for Gaia, since it may have been the event that led to the abrupt break in the fossil record that geologists call the Cretaceous/Tertiary (K/T) boundary (see illustration overleaf). The break indicates that about 65 million years ago the Earth experienced a nasty turn. It may have been the devastation of the planetesimal impact, or the volcanism that followed, but there is no doubt that the environment was so disturbed that most of the marine life died. This led to a disturbance of temperature regulation from which the system took a considerable period to recover.

In a *Nature* article, Jack Wolfe describes the distribution of species of fossil plants found in various parts of the Earth before and after the (K/T) event. The shift of the latitudinal ranges of the plants provided impressive evidence of a rise in the average temperature of the whole Earth of 10°C after the event. The evidence further suggested that this temperature rise persisted for between half and one million years.

I find this a most significant observation for Gaia. If confirmed it justifies my belief that the carbon dioxide of the air is kept low by the continuous pumping affect of living organisms in the soil and the oceans. The K/T event killed off as much as 60 per cent of the organisms and reduced the rate of pumping down of carbon dioxide. Not only this, but large quantities of carbon dioxide would have come from the heating to incandescence of crustal rocks and ocean water, and from the fermentation and oxidization of the 60 per cent of the biomass that died. The rise in methane and carbon dioxide would have raised the Earth's temperature. The destruction of marine algae may also have reduced DMS production and hence cloud cover but these are speculations about what could have happened; the exact course of events at and following the impact are still unknown.

The paleobotanical observations suggest that it took the Earth system about one million years to recover from this impact.

The cycle of carbon dioxide involving the rock weathering sink is long and slow. But not so slow that it would take a million years to be resorbed. The most awful thought about what we humans are now doing to the world is the length of time the injuries will take to heal.

A complex system

The models of climate regulation are, by their nature, very much simplified. On the real Earth the regulation of the climate (if it happens) is an amazingly complex affair. Here are some of the as yet little understood factors that may be involved in Gaia's temperature control.

Craters on the Canadian Shield

Planetesimals are chunks of rock moving in irregular orbits through the solar system that may in time collide with one of the planets. They can be up to 16 kilometres in diameter (if one of these was resting on flat ground it would be nearly twice as high as Mount Everest). The pock-marked faces of Mars and the Moon, with no rainfall or life to wear away the scars, show how prevalent these collisions are.

On Earth, the craters are not usually visible at ground level, having long since been filled in and eroded away, but their presence can be seen in space photographs like the one below right. They can also be revealed by geological surveys. Made from such data collected by the Canadian Government, the map (right) shows the large craters in the old, hard rocks of the Canadian Shield. The Canadian Government's interest was largely economic: planetesimals are often made of nickel–iron alloy and their craters can yield a rich supply. The great mines at Sudbury, Ontario are an example.

Energy of impact

Planetesimals travel through space at up to 50,000km/h, 30 times as fast as a rifle bullet. A bullet at that speed would have 900 times its usual energy, because energy is proportional to the square of the speed. Its impact at 50,000km/h would release energy equivalent to the detonation of 40g of high explosive, enough to fragment you completely. The impact of a 16-kilometre diameter planetesimal at 50,000km/h releases a thousand times more energy than the simultaneous detonation of all the nuclear weapons in the world.

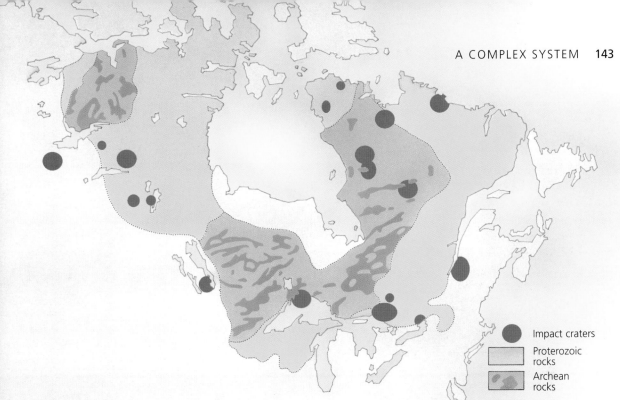

Impact craters

Proterozoic rocks

Archean rocks

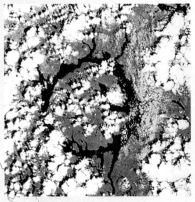

Manicougan Crater

The infrared space photograph (above) shows the Manicougan Crater in Quebec, Canada. The crater has a ring-shaped lake around its rim and was formed by a meteorite impact.

The Canadian Shield

The landsat photograph (right) of north Saskatchewan and Northwest Territory, Canada, shows the Canadian Shield, with its thousands of small lakes.

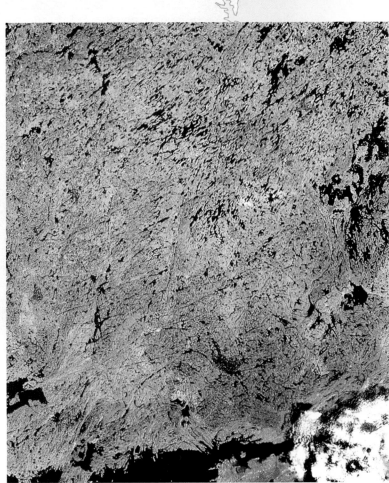

CASE HISTORY The Planetary Gunshot Wound and the Dinosaurs

One of the liveliest events to relieve the gloomy solemnity of modern science was the discovery by the Alvarez family in the Italian town of Gubbio of a great enrichment of the precious metal iridium at what geologists call the Cretaceous/Tertiary (K/T) boundary.

Now the K/T boundary marks the point where a long history of rock formation by the slow deposition of the calcium carbonate shells of marine algae suddenly ceases. The layers of rock more recent than this are of sands and mud devoid of the abundant shells of the Cretaceous. This is consistent with the idea of a great disaster, where almost all of the organisms in the sea above the site of deposition were killed at once. The Alvarez team suggested, I think most reasonably, that the presence of iridium and similar elements in the rocks of the boundary indicates that the disaster in question was a planetesimal impact. Their argument is that iridium is rare on Earth's surface, while meteorites and rocks from space are often enriched in iridium and similar elements. It would take a chunk of rock from space, rich in iridium, to raise the level in a thin layer all around the Earth. The vast energy of the impact would have vapourized and distributed the elements of the planetesimal all around the world.

What caused the excitement was that the boundary occurs more or less at the time when the dinosaurs became, if not extinct, at least much less conspicuous than they were before.

So it seemed natural to associate this great injury 65 million years ago with the demise of the dinosaurs. But the innocent geophysicists and geochemists who raised these ideas did not reckon with the rage of the paleontologists who resented the suggestion that their pets had been suddenly destroyed as the result of a planetary gunshot wound. They were sure that the extinction of the dinosaurs had occurred gradually as a result of evolution by natural selection, and that no sudden event was involved.

Scientists are beginning to realize that both explanations could be right. Just as a bullet does not always kill instantly but can set in train a series of physiological events that kill many years later, so an impact can have consequences that reverberate through all the geophysiological systems with their widely varying response times. A considerable part of the damage caused by an impact to Gaia could have been delayed for over a million years.

The geologist, Michael Rampino and his colleagues raised an intriguing notion concerning the great volcanic outbursts that occurred about a million years after the dates of the planetesimal impacts. During these outbursts, thousands of square kilometres of land were covered by glowing lava – enough to disturb the heat balance of the whole planet. Their suggestion was that the shock wave of the impact as it moved in the body of the Earth became intensified enough to melt a region of the plastic rocks deep beneath the Earth's crust. This pool of molten rock took a million years to move to near the surface, where it manifested itself as a huge series of volcanic eruptions, which must have affected the Earth's atmosphere and climate profoundly. The vast, lava-formed plateau of Southern India known as the Deccan Traps (see right), and the lava fields of Idaho in the USA, are results of these extraordinary events. Such ancient lava flows, covering tens of thousands of square kilometres, with a succession of flows forming plateaus, can be found all over the planet, both beneath the oceans and on land.

Not only this but the destruction of the marine and surface life by the impact must have led to a large increase of carbon dioxide in the air, probably much larger than the one we shall soon experience if fossil fuel burning continues, and this increase in a greenhouse gas would have further raised the temperature.

Analysis of fossil evidence now suggests that there was a 10°C rise of global temperature, four times larger than the rise from the last ice age until now. This would have vastly stressed the ecosystems of those times. Perhaps the dinosaurs died not directly from the impact, but because they could not compete in the changed and hotter world, which lasted for a half to one million years.

The Deccan Traps

The vast basalt plateau of Southern India known as the Deccan Traps was formed by an outpouring of volcanic lava about 65 million years ago. In Mysore (above) the high plateau rises up above the fertile valleys since created across it by erosion.

Sandstone (Tertiary)

Iridium (boundary layer)

Chalk (Cretaceous)

The Cretaceous/Tertiary (K/T) Boundary

About 65 million years ago the annual rain of coccolithophore shells to the sea bed ceased, and was replaced by a rain of inert mud and sand (p. 142). The rocks formed from these layers of sediment show a clear boundary between the chalk laid down during the Cretaceous and the mud and sand from the Tertiary period. The coin gives an idea of the scale.

The greenhouse gases

I have shown how life is involved in all stages of the great carbon cycle that sustains current levels of the greenhouse gas, carbon dioxide, in the air. Life is involved, too, in the other principal greenhouse gases – methane produced by the methanogens, nitrous oxide produced by denitrifying bacteria and, of course, water vapour. Life is involved in evapotranspiration of water, in cloud formation, and in the flow of water through soil. Without the presence of life on Earth, as we have seen, water could have left this planet. So to understand how the greenhouse works – and we urgently need to understand – we have to take into account all of life.

The other thing to know about the gaseous greenhouse is that it isn't quite the simple story I have just told. One complication comes from the fact that by far the most important of the greenhouse gases is water vapour. It's the most abundant; there are at least 10,000 ppm of it, so it overshadows all of the other greenhouse gases in abundance and is the source of most of the gaseous heating. But water vapour, as you well know, condenses out as rain or snow if the air gets cold. The concentration of water vapour in the air therefore depends on the air temperature. Water vapour acts, in a way, to amplify the effect of the other greenhouse gases. The warming by carbon dioxide, for example, is mostly due to the water vapour that the warmer air can hold. Gases that absorb the infrared and don't condense out are always there and their presence lifts water vapour into the air. That water vapour then absorbs more infrared and so the process goes on until an equilibrium amount of water vapour is kept in the air by the presence of carbon dioxide and the other greenhouse gases.

The albedo effect

Seen from space, our planet is a sapphire blue sphere, capped by brilliant fields of polar ice, and swathed in white cloud. The darker oceans and forests absorb heat from the Sun. The snow, ice, and clouds, through their light colour, contribute to the cooling albedo effect by reflecting the heat from the Sun back into space. Could living organisms be involved in regulating albedo, and so in planetary temperature control?

In many ways the Earth is like Daisyworld (see p. 67), tending to resist change away from comfortable temperatures in the 14 to 20°C range. Light and dark daisies are insufficiently prevalent to affect temperature directly, but large ecosystems exist that do affect climate.

On the land surfaces in the temperate regions the coniferous forests act like dark daisies, with the effect of extending the growing season deeper into the winter. Dark stands of conifers whose shape has evolved to shed snow absorb sunlight and warm the region. Richard Betts and his colleagues showed in a *Nature*

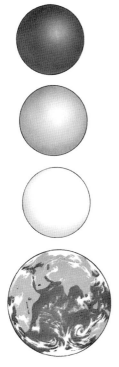

Black planet
Albedo 0.0

Grey planet
Albedo 0.5

White planet
Albedo 1.0

Planetary
albedo
topography

The albedo effect

Albedo is an astronomer's term for the depth of colour of a planet, that is, its lightness or darkness, and hence its reflectivity. A black planet, reflecting no light, has an albedo of 0.0; a grey planet, reflecting half the light incident on it, has albedo 0.5; and a white one, reflecting all incident light, has albedo 1.0.

Put most simply, the surface temperature of a planet depends on the balance between the heat it receives from its sun, and the heat it returns to space. On Earth this equation is complicated by the greenhouse effect. It is also strongly influenced by albedo, which varies with the planet's topography (see diagram). Light areas such as polar caps, snow, or clouds can reflect 70 or 80 per cent of incoming sunlight back to space. They have high albedo. Dark areas such as belts of forest, or oceans, have low albedo; they absorb the Sun's heat, and then radiate it back in the infrared, in which case it can be retained by the greenhouse gases. The Earth's albedo changes seasonally with ice and cloud cover and the growth of vegetation.

paper in 1998 that the boreal forests of Siberia and Canada have a marked affect on the temperatures predicted by global models that include them.

In the tropics there are, or used to be, the rainforests. These vast ecosystems, before their removal by farmers and ranchers, covered most of the tropical land masses. The forests require abundant rain for their existence, and they make it by evapotranspiring huge quantities of water vapor that sustain rain clouds above them. These clouds also act like white daisies, reflecting the sunlight, and keeping the tropical regions cool.

Over the open ocean are algal ecosystems that exist in gigantic blooms covering areas thousands of kilometres in diameter. Production of DMS (dimethyl sulphide) by the algae (see p. 124) adds this gas to the air where it oxidizes and forms microdroplets of sulphuric and methane sulphonic acids, which are the nuclei without which there would be fewer and less dense clouds. Oceans cover about two thirds of the Earth's surface and are dark blue. Anything that affects cloud cover over them can powerfully affect the climate of the Earth.

Biological feedback

The cloud–algae link has become an exciting scientific area. Two groups of French glaciologists (C Lorius, J Jouzel, D Raynaud, J Hansen, and H Le Treut) reported in *Nature* in 1990 their discovery of sulphuric and methane sulphonic acids in Antarctic ice cores, going from the present to 30,000 years ago. Their data show a strong inverse correlation between global temperatures and the deposition in the ice of these acids: as the temperature decreases the deposition of the acids increases. Sulphuric acid has several natural sources, but methane sulphonic acid is unequivocally the atmospheric oxidation product of DMS. There was a two to five times larger deposition of this substance during the Ice Age, and it seems probable that this was due to a greater output from the ocean ecosystems, probably a consequence of a greater supply of nutrients to the cooler surface waters which were less stratified than now. If confirmed, this suggests that algal-increased cloud cover and low carbon dioxide operated in synchrony as part of a geophysical process to restore the Earth from its fevered inter-glacial state to the cool comfort of an ice age. Remember, Gaia's preference may not be the same as ours. These speculations received theoretical confirmation when Lee Kump and I modelled the effects of global cooling on algae and land plants and reported our findings in a *Nature* paper in 1996. The cloud–algal theory is becoming part of conventional scientific wisdom and must now employ hundreds of scientists world wide in its investigation. I hope that it will not be forgotten that it was inspired by Gaia theory. My friend and colleague Robert Charlson thinks that the emissions of DMS by algae in the natural world is worth about 10°C of cooling. The clouds it seems are our most effective white daisies.

A crisis of climate control

When I was visiting NCAR (National Center for Atmospheric Research, Boulder, Colorado, USA) in 1986 at the time of its 25th anniversary, I enjoyed some lively debates with Stephen Schneider, one of the best climatologists. I recall him saying, "Jim, how can you possibly believe in Gaia when there are ice ages? If there were a Gaia it would stop them from happening."

As always in science the interesting developments come from good and fair criticisms. Stephen Schneider's comment led me to consider the possibility that an ice age, which lasts about 100,000 years, may now be the normal comfortable state of the Earth and that the warm periods between are short-lived but recurrent fevers of Gaia.

The Earth, as a home for life, is now 3.8 billion years old and with a lifespan not more than 4.5 billion years. Not surprisingly, it shows signs of ageing, and some systems are not working as well as they might. A younger planet might not so readily flip between the cold of an ice age and the heat of an interglacial as appears now. Indeed a planetary physician would see the present series of glaciations interspersed with brief warm spells as evidence suggesting a crisis of climate control, a result possibly of approaching senescence in an ageing system.

The physician would know that ever since the Earth formed the heat it has received from the Sun has increased. Over her 3.8 billion years of life, Gaia's homeostatic systems have kept the temperature comfortable despite this warming Sun by pumping down carbon dioxide. The abundance of this gas has declined by a factor of five hundred. It fell from its pre-life level of about 10 per cent to about 0.1 per cent after the start of life; and has reduced to a level of only 180 parts per million in the recent "normal" glacial state. This decline has kept the gaseous greenhouse right for an increasing solar output and for a constantly changing set of living organisms. But in the last few tens of millions of years the solar output has reached a level where it is becoming increasingly difficult for the carbon dioxide pumping system to operate. To keep cool when the solar output is as high as now requires efficient pumping by the system so that a carbon dioxide level below 200 parts per million is sustained. This is near the lower limit at which plants (even the more recent, carbon dioxide efficient type, see right) can grow. At this low level plant growth declines, and efficient pumping might then require a larger land area to support this less effective plant life – just such a larger land area as is available during the ice ages (see p. 151). (The restriction applies much less to ocean life where carbon dioxide is abundant because the ocean is largely a solution of bicarbonate.)

Plant life and carbon dioxide

As the sun increases its heat output, Gaia has to maintain ever lower levels of carbon dioxide to keep comfortably cool. Current levels of 180 ppm in glaciations and 280 ppm during interglacials are close to the lower limit at which plant life can thrive. Not surprisingly, this has created an evolutionary pressure for new types of plant able to grow at lower CO_2 levels. About 10 million years ago such a new type did evolve – the C4 plants, as distinct from the C3 plants that had dominated until then. (C3 and C4 refer to different types of carbon metabolism: the C4 plants are better able to photosynthesize with lower available CO_2). The new type includes most of the grasses – a highly successful group. The success of the C4 plants can stave off, but not prevent, the ultimate failure of the present cooling mechanism. Within 100 million years, the Sun will be warm enough to require zero carbon dioxide to maintain present temperature norms. The system will then have to shift to a new set of parameters.

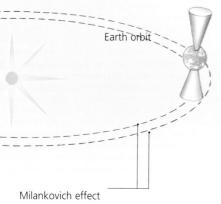

Earth orbit

Milankovich effect

The Milankovich effect

The Yugoslav meteorologist Milutin Milankovich was the first to observe that variations in the orbit and inclination of the Earth in its motions around the Sun led to periodic increases and decreases in the heat received from the Sun and also to changes in the relative heating of the northern and southern hemispheres in winter and summer. Just now, for example, northern winters are warmer and southern winters cooler than average. He proposed a link between glaciations and these astrophysical effects. The magnitude of these variations in solar heat received by the Earth is not enough in itself to account for the flip from glacial to interglacial, but the synchronicity between the orbital cycle and the glacial one does suggest that the Milankovich effect is the trigger.

That the system is stressed is evidenced by the fact that even the small additional flux of heat that occurs when the Milankovich effect (periodic changes of orbit, see illustration) brings the Earth closer to the Sun, is enough to destabilize the healthy glacial state and bring on the fever of an interglacial.

The present oscillations to and from a glacial state are recent: up until 2 million years ago, the climate was much more constant. Glaciations have occurred in the Earth's past history, but long periods free of ice have tended to be the rule. You should keep in mind that while a warm spell when the Sun is hot could be an illness, when the Sun was cool and when oxygen first appeared, a cold spell could have been a disease. (Like the iron in Chapter 3, Gaia would normally regulate on the warm side in cool conditions, and on the cool side in warm ones.)

The Sun grows ever hotter, for that is the nature of stars. When will it be so hot that the present carbon dioxide pumping method of temperature regulation will fail for the contemporary biota? The answer must be before 100 million years have elapsed, for by then the abundance of carbon dioxide in the air needed to keep the temperature at present levels will be approaching zero.

Failure of the present system will not mean the death of the planet, only a change in the method of regulation and a different biota living at a new stable hotter state. We have seen that Gaia already has an additional cooling mechanism in the generation of clouds by algal emissions of sulphur gases (see pp. 124–5). What we do know is that the continuation of Gaia does not require the present very low mean temperature of the Earth, 11.5°C in the glaciations and 14°C in the interglacials. In fact, the optimum temperature for plant growth (but not for humans) is nearer 25°C. If an entirely new system evolved operating at a temperature of around 25°C, the pump down of carbon dioxide could resume and continue for a further billion years. This guess at the probable life span of Gaia made in 1990 has received general support from modellers including the American geochemists, Kasting and Caldiera and by the German scientists Shellnhuber, von Bloh and their colleagues.

If the geophysiological analysis of the ice ages is correct and the present interglacial is a fevered state, then the pollution of the air with greenhouse gases and the clearance of the forests to make more farm land to feed people are more serious now than they might have been during a glaciation, when the Earth was somewhat healthier and less perturbable. Forests evaporate water just as does a sweating man; both actions serve to cool. The man is cooled by the evaporation itself, the forest by the clouds that form above and reflect the hot rays of the Sun back to space. Adding greenhouse gases to the air is like piling blankets on to an overheated patient. By our pollution and by our destruction of natural ecosystems we add

CASE HISTORY The Relapsing Fever

Are the ice ages, which last nearly 100,000 years, the healthy state of an elderly planet, and the brief warm spells the fevered state of a planetary illness? Of course to us, the intelligent microbes of the Earth, the present warm state seems just right – just as, if it could think, a pneumococcus might feel about a pneumonia in one of us. We are so set in thinking of the Earth in human terms that we imagine an ice age to be a planetary catastrophe. True, for human civilizations spread over the northern temperate zones, an ice age would be disastrous. Much of Europe north of the Alps, and America north of 45 degrees latitude would be under thousands of metres of ice. Bad for humans – but marvellous for the planet.

For during a glaciation, the sea level would be more than a hundred metres lower than now, and an area of sea floor equal in size to the continent of Africa would become land surface. The bulk of this new land would be in the tropics. Evidence from the ice cores of Greenland and Antarctica shows that during the last glaciation the output of DMS from the oceans was five times as great as now, and that the carbon dioxide of the air was reduced to 180 ppm – about 60 per cent of the level before humans began to disturb it. The increase of DMS is evidence of a more vigorous ocean life, and the reduction of carbon dioxide implies increased pumping by a larger and more active biota. However humans may feel about an ice age, it seems that Gaia likes it cold.

So what causes the fever? The short-term answer is geophysical. The Milankovich effect (see p. 149) certainly explains exactly the timing of the interglacial warm periods – they coincide with one of the times when the Earth is nearer the Sun. But it is an incomplete explanation on two counts.

First, the change in temperature is larger than would be expected from the small change in solar heating, and second, it takes three successive Milankovich warming events to break the long spell of glacial cold.

There are geophysiological grounds for believing that the warm spells are, like fevers in humans, the indication of an underlying pathology. If the Earth's temperature regulation were in good order, the minor wobbles of its orbit would never have an effect as drastic as the jump from the glacial to the interglacial state. The recurrent warm spells now experienced are the pathology of a form of planetary senescence. The glacial state is unstable, as any significant warming and melting of ice would lead to the inundation of land surfaces and the killing off of plants that otherwise would be serving to pump down carbon dioxide; moreover there would be a surge of methane from the inundated plant life. The increase in atmospheric carbon dioxide and methane would enhance the warming and the inundation, and soon a runaway positive feedback on warming would take place. Physics and physiology would conspire together to hasten the warming and the melting of the ice, so that the transition from glacial to interglacial might take only a few tens of years. After a period of interglacial warmth there would be a slow recovery , with the system growing cooler towards a

new glacial period. As the ice age continued, it would slowly become more sensitive to perturbation.

This reasoning can explain why the Milankovich heating at 40,000-year intervals does not trigger an interglacial, for it comes at a time when the ice age is "younger" and the system is still relatively stable and insensitive to small perturbations.

The relapsing fever of the recurrent interglacials has afflicted the Earth for over 2 million years. Past episodes have not been seriously disabling and recovery has soon taken place. But the fever may be a warning of a more serious disease – failure of the carbon dioxide pumping system. And a planetary physician would note that in the ageing state of the CO_2 climate regulation system the sudden appearance of *Homo sapiens*, with a tendency to increase the gaseous greenhouse, was a complication that gave cause for concern.

The healthy ice age

How could there have been more life
during the ice ages? Wouldn't the ice
sheets have left less room for life? Strange
as it may seem, it could have been quite
the reverse. The withholding of water
from the oceans to form land glaciers
lowered the sea level by more than 100
metres. As this map shows, these lower
sea levels would have exposed large areas
of land on the continental shelves – much
of it in the near Equatorial regions of
South East Asia and the Pacific. The extra
land area could have been as large as
Africa today, and could have included
areas rich in land life.

A failing system?

As a self-regulating homeostatic system is
stretched to the limit of its capacity to
regulate, it grows unstable; the instability
can reveal itself as an oscillation between
two limits, like the wobble of a gyroscope
before it falls. The present
glacial/interglacial oscillation is illustrated
in the graph below, which plots
temperature against time before the
present. It behaves as if it were a free-
running physiological cycle with a period
of, say, 150,000 years, triggered at
100,000-year intervals by the Milankovich
pulse of heat.
     The temperature difference between a
glaciation and a normal interglacial is
about 5°C and that between normal and a
fever in one of us about 3°C. These are
variations within physiologically acceptable
bounds and do not in themselves imply a
failure of temperature regulation.

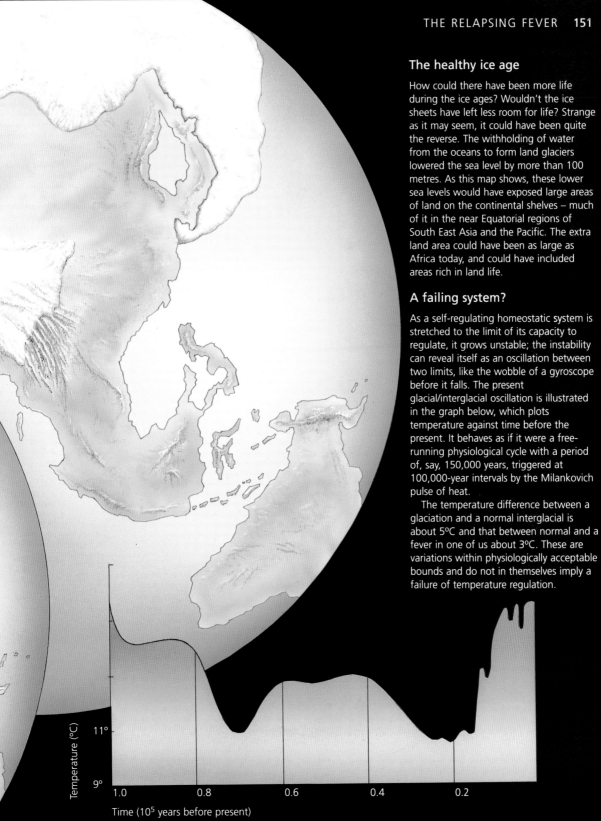

Temperature (°C)

11°

9°

1.0   0.8   0.6   0.4   0.2

Time (10⁵ years before present)

to the burdens of an already fevered planet. As in a fever the Earth's present state is one of positive feedback on temperature rise. The systems of cloud cover and greenhouse gas regulation which normally would act to regulate now tend to amplify any change in temperature. Among the unpleasant surprises in store if we continue in this way could be a premature jump to another stable system state, most probably hotter and profoundly uncomfortable for the current civilizations.

CHAPTER EIGHT

The people plague

This book has been about the Earth as a system, Gaia, and I have assumed that it is sufficiently like a living organism to be subject to illness or injury. I have described the anatomy, biochemistry, and physiology of this system, and used the metaphor of planetary medicine to trace its medical history from birth, through childhood, up until now. And I have shown, in case histories, how Gaia is self-healing and has coped with many accidents and diseases.

Now I want to illustrate the present state of the Earth by the view seen through the eyes of a planetary physician. Pretend, once more, that the Earth is a patient visiting a cosmic clinic: it's time for a check-up. The practitioner at the clinic sees a comfortable elderly planet, apparently in a good state of health. But the reports in from the pathologist and dermatologist are of unusual physical signs. Atmospheric CO_2 and methane are above the patient's normal range, and there is a suspicion of fever. Some skin damage is apparent – the land surface shows a number of bare patches. Most revealing are certain abnormal chemicals in the air – CFCs, substances that are never made by the natural chemistry of living organisms. To the practitioner, atmospheric CFCs with an abundance of one part per billion suggest the presence of a highly organized social species, one that has an advanced chemical industry.

The physician knows that old planets bearing life have the capacity to evolve intelligent species, and that the presence of such organisms indicates a potential, if not an actual, illness – a condition which may be damaging, but is almost never fatal. Something that might even be turned to the patient's advantage.

Humans on the Earth behave in some ways like a pathogenic microorganism. We have grown in numbers and in disturbance to Gaia, to the point where our presence is perceptibly disabling, like a disease. As in human diseases, there are four possible outcomes: destruction of the invading disease organisms; chronic infection; destruction of the host; or symbiosis – a lasting relationship of mutual benefit to the host and invader.

If microorganisms were sentient, they would realize that in the long term their future lay in working to attain the fourth state, that of symbiosis – a binding contract, where the invader is restricted to a region or role and there supported, protected, and nourished, in return for some service to the host. Within each of us as individuals there are many such contracts with microorganisms. At the deepest level is Lynn Margulis's endosymbiosis: a state in which the once free prokaryotic consumers are now forever domesticated as the mitochondria of our cells. For them and us the contract is wholly binding, and determines the very existence of both. Other familiar examples are the communities of bacteria that live in our mouths and our guts, organisms that may, long ago, have been unwelcome guests, but now defend their territory and ours from less desirable occupants, and do things to our food that makes it more digestible and nutritious.

The precedent in nature for enduring contracts of this kind is so strong that as an intelligent species we already have the map of the way ahead. All we seem to need is the will to travel by that route. There are however some inherent properties of humans that make it difficult for us to act sensibly and achieve symbiosis within Gaia.

As a species, we live neither as free and independent individuals, nor as completely integrated social organisms like the bees. Rather, we live tribally; and our tribal behaviour is all too often far below the standard of the best among us. Intelligent we may be as individuals; but as social collectives we behave churlishly and with ignorance. I think that our inability to live in harmony with one another and with the Earth comes from this disparity – from the gap between the power of our human collectives to act, and the feeble intelligence that directs that action.

As individual humans, or as small groups hunting and gathering, we once lived in symbiosis with our planet. When we began using fire, tools, and agriculture, we became more dependent upon each other socially, and also more powerful and numerous. We had the potential to sustain our own environment at the expense of the Earth; to break our contract with Gaia. At first the breaches were mild, just the wearing of clothing and the building of houses. Then we began to herd our food prey, cattle, and to grow our favourite food plants. But right up until the beginning of this century none of this, nor the industrial civilisations that had evolved, were significant in themselves to the Earth. The danger lay in the potential for further growth and development. Now the consequences of that growth in our numbers and development of our capacity to displace the rest of planetary life, threaten both us and our planet.

The first stage of the growth of populations of successful organisms – the stage when food is sufficient – often takes the form that mathematicians call exponential. If you ask what exponential means, they answer: "Quite simple; exponential growth means that

The four outcomes of disease

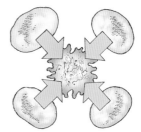

1 Invading microorganisms, are destroyed by the host's defences.

2 The host and parasite settle down to a long war of attrition. The state of chronic infection.

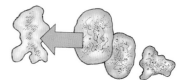

3 The host is destroyed by the parasite, which then dies also.

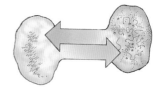

4 The host and microorganism enter a new relationship of symbiosis, a partnership based on mutual benefit.

What will be the outcome of Gaia's uncomfortable plague of people? The four possible scenarios are listed above. The last, symbiosis, is obviously desirable. As intelligent microbes, we have the advantage of knowing the risks of failure, and the lasting benefits of symbiosis. But will we achieve it?

the rate of increase is proportional to the numbers present". Simple it may be, but few comprehend its awesome consequences.

To understand what exponential growth is really like, imagine a pond on whose surface a water lily is growing and spreading so as to double its leaf area daily. It has taken the lily 19 days to cover half of the pond's surface with leaves. How long will it be before the leaf area doubles again so that the pond is entirely covered? Does the right answer come instantly to mind? Not another 19 days, but *one day*.

And when the lily pond is covered, what then? Growth in any natural population never follows an exponential course indefinitely. It is easy to calculate that with exponential population growth the mass of, say, bacteria, or of house flies, would eventually exceed that of the Earth itself, if unchecked. The real relationship is closer to what is called logistic: exponential at first, then slowing down to zero growth at a steady state. In practical terms this can mean that the death rate rises to meet the birth rate, as environmental constraints on the population come into play.

We humans must be close to that 19-day period of lily growth so far as our coverage of the Earth's surface with crops, livestock, industries, and cities is concerned. Our exponential growth will soon cease through its own contradictions. As the 19th-century economist Thomas Malthus foresaw, the human population is outstripping its food supply and environment and must in the end be curtailed – by famine, disease, war, or natural catastrophe. Malthus' predictions are coming true at last. The statement "There is no pollution but people" carries an awful truth.

Pollution is always about quantity. In the natural state there is no pollution. The dung of a grazing or browsing animal does not foul the Earth and stop the plants from growing; it feeds and nourishes them. But the dung of a 100 cattle kept by a greedy farmer in too small a field is a real pollution and destroys the grass they feed on.

None of the environmental agonies now confronting us – the destruction of the tropical forests; the degrading of land and seas; the looming threat of global warming; ozone depletion and acid rain – would be a perceptible problem at a global population of 50 million. Even at a billion people, these pollutions would probably be containable. But at our present numbers – more than six billion – and present way of living, they are insupportable. If unchecked, they will kill a great many of us and other species, and change the planet irreversibly.

As a vast collective, the human species is now so numerous as to constitute a serious planetary malady. Gaia is suffering from *Disseminated Primatemia*, a plague of people. In human medicine death rarely comes from a major illness itself. The unfortunate person paralyzed by injury, or by infection with a virus disease such as polio, does not die of paralysis but from pneumonia, or urinary tract infections, brought on by the disturbance of natural function due to the paralysis. Few are they that die of old age and nothing else. Yet ageing is the main cause of death. In the same way if the main disease of the Earth is the superabundance of people at the present way of living, harm will come, not just from their presence, but from the disturbance of the Earth's natural function by what they do.

Let us now look at some of these disturbances and at current human attitudes to their significance as environmental threats, and compare these attitudes with the view of a planetary physician – taking each threat as if it were the case history of a specific malady afflicting Gaia, but knowing that there would be no such malady if there were fewer people, or if they lived in harmony with the Earth.

Agriculture and deforestation

I think that by far the greatest damage we do to the Earth, and thus by far the greatest threat to our own survival, comes from agriculture. We shall soon have taken away more then two thirds of Gaia's natural terrestrial ecosystems and replaced them with agricultural systems. When we replace natural forest with food crops or cattle farms we diminish the ability of the land surface to control its own climate and chemistry. Our human substitute ecosystems are there just for our own or the farmer's short-term benefit. When we farm, unless we do so very sensitively, we are evading our contractual obligation to Gaia – and most farming, especially agribusiness, is grossly insensitive. In a way, when we chose to be farmers, we broke our links with Gaia and fell from paradise.

The great ecosystems of the Earth, the forests, the marshes, the shallow seas and lakes, the algal blooms of the oceans, all these and others are part of the regulatory system of Gaia. As with all living systems there is a great deal of over-provision and much can be destroyed or replaced with "productive" (in human terms) but inefficient (in Gaia's terms) ecosystems without too much loss. But that over-provision of Gaia is no luxury; it is there to cope with abnormal stresses. You or I can manage quite well with only one kidney. It would not be wise though to remove and sell one of your kidneys if you were due to face the stress of dehydration crossing a desert on foot in the heat of summer.

As polluters, we alter the atmosphere, waters and soils of the Earth, and so increase the stress to which the natural ecosystems are subject. But as farmers we do still greater harm, by clearing the land and so reducing the capacity of the whole system to deal with stress. Consider a humid tropical forest. It will be part of the territory of some nation state that considers it to be right to cut it down, to sell the timber that is valuable and burn the rest so as to leave land that can be farmed. Until recently we might have thought: "Why not? It is their human right to exploit the land that is theirs."

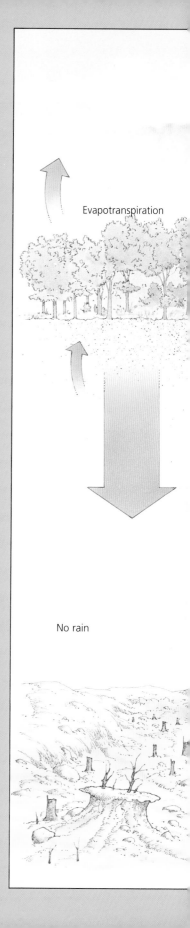

Evapotranspiration

No rain

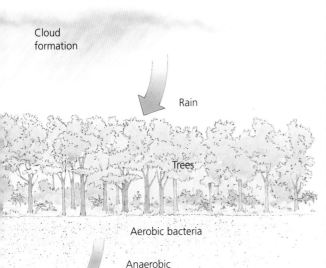

Cloud formation

Rain

Trees

Aerobic bacteria

Anaerobic bacteria

The loop: trees, rain, and soil

The wet and cloudy tropics are not a given state of the Earth. The trees themselves keep things this way, by evapotranspiring huge volumes of water through their leaves. The rising vapour condenses to form clouds, the rain falls, the trees grow, and their roots bind the shallow soil and leaf litter, where nutrients are rapidly recycled by bacteria.

Opening the loop

Take away the trees, and the rain will cease. Trees and rain go together; without one there cannot be the other. Without rain, the soil will begin to die, too, as the bacterial ecosystems that sustain it are exposed to harsh conditions and erosion. The forest will not return – and the land with turn to scrub, or desert.

Soil erosion

CASE HISTORY Exfoliation

To a planetary physician, by far the most dangerous malady afflicting the Earth is that of exfoliation – destruction of its living skin. In human medicine the loss of skin from whatever cause is a serious threat to life: the loss of more than 70 per cent of the skin by burning is usually fatal. To denude the Earth of its forests and other natural ecosystems and of its soils is like burning the skin of a human. And we shall soon have destroyed or replaced with inefficient farmlands 70 per cent of the Earth's natural land surface cover. Our annual toll of tropical forests nearly doubled between 1979 and 1989, while each year sees almost 21 million hectares of land degraded to useless scrub or desert, and 24 billion tons of topsoil washed or blown away through erosion as a result of our efforts.

Soon we will have removed 65 per cent of all the forests of the tropics. When more than 70 per cent of an ecosystem is lost, the remainder may be unable to sustain the environment needed for its own survival. So the rest of the trees may then die too.

The great forests of the tropics are part of Gaia's skin, and like skin they sweat to keep us cool. The tropics are warm, humid, and rainy, an ideal environment for trees. But few seem aware that the trees themselves keep things this way. The trees of the forest bring the rain and the rain allows the forest to grow. Take away the trees and the loop is opened (see left); the rain ceases and the region turns to scrub or desert. Trees and rain go together as a single system. Without one there cannot be the other.

The Harrapan Desert

Perhaps the best known example of the pathology of forest loss is Harrapan in Western Pakistan. The region was once abundantly forested and enjoyed an adequate rainfall during the monsoon season. It was a fine example of a self-sustaining forest ecosystem. The forest was gradually cleared by peasant farmers who kept cattle and goats that grazed on the scrub and grass that replaced the forest trees. The rainfall was sustained over the region until rather more than half of the forests had been cleared. But after that the region became arid and the remaining forest decayed. The region is now so dry that as a semi-desert it can support only a fraction of the people and other organisms that once were there.

But now, dimly at first, we begin to realize that if we and they allow such exploitation to continue, all of the Earth is affected. Furthermore, the expectation of a profitable farming economy based on forest clearance is impermanent. Take the trees away and the rain goes with them. The climate of the region changes – it grows hotter and drier. The nation may find it can no longer feed or sustain its people, the world grows slightly hotter because of what they did, and Gaia has lost some more of its capacity to resist further change.

The effects of forest clearance will probably be the first gigantic global disaster to greet us within the next decade or two. Despite all the world protests, we are still destroying the tropical forests at a ruthless pace. Yet scientists and campaigners in the First World and in the tropics still try to plead for the preservation of these forests on the feeble and human-centred grounds that they are home to rare species of animals and plants, and especially to plants containing drugs that could cure cancer or other frightening human diseases. This may be so. But they offer so much more. Through their capacity to evaporate vast volumes of water vapour the forests serve to keep their region cool and moist, by wearing a white sunshade of reflecting clouds and bringing the rain that sustains them. More even than this, the great forests of the tropics are part of the cooling and air-conditioning system of the whole Earth.

Maybe you think that the forests are so vast that it will take decades to clear them significantly. If you do, you could be wrong. The rate of clearance worldwide is now so great that if it continues, then by 2010, more than 65 per cent of the forests will have gone. Numerical models based on Gaia theory, and the experience of past civilizations, both predict that once more than this proportion of a self-regulating ecosystem dies, then it can no longer sustain its climate and total collapse takes place. If we delay our decision to stop felling the trees until 70 per cent are gone it might be too late; the rest would die anyway. If we let deforestation continue we may soon reach the day when at least a billion people are living in these once-forested regions, but in a hostile climate, hot and arid – an unprecedented human political problem as threatening as a major thermonuclear war. If it should happen, we shall be faced with the problems of the South Saharan drought multiplied a hundredfold. We are now failing to support a few millions in need in the Sahara. How could we possibly sustain the billion? Do we let them die? Can that number of refugees be accommodated?

And if we continue to denude the land, waste the soil, and demand more and more of the Earth for our crops and livestock, we will rapidly approach the critical level of degradation of *all* Gaia's natural terrestrial ecosystems – the level where the whole system starts to collapse.

Acid rain

Ever since oxygen became the dominant atmospheric gas, rain has been acid. It may well have been acid always, but there is just a chance that once in a while in ancient pre-oxygen times, it was neutral or slightly alkaline.

If the rain has been acid for over 2.5 billion years why is there a problem, and what is all the fuss about? The problem is the extra acidity of rain falling from polluted air; this is damaging to trees and sometimes also to the natural ecosystems of rivers and lakes and should be curtailed. The fuss, however, is often more to do with political or tribal concerns than with the scientific truth about acid rain or its environmental consequences. The acid rain controversy flourishes with great public support. Complicated questions such as "What is an acid?" or "How acid is the rain?" stir little interest compared with disputes about who is to blame.

If you think I am being frivolous, let me tell you of my experiences in the European battle over acid rain. At a meeting early in the 1980s between the scientific establishments of the Scandinavian countries and of the UK, one of the spokesmen for the Scandinavians said "Gentlemen. The object of our commission is to prove that Britain is responsible for the acid that falls on Scandinavia." The agenda before the meeting allowed no confusion of the issue with new facts. It went straight to the point: find the evidence that would put the blame on the Brits.

So what were the facts? It is true that about 15 per cent of the acid in the rain falling on Scandinavia comes from combustion sources in the UK. It is also true that lakes and forests in Scandinavia suffer from the acidity of their waters. Fish die and trees are damaged. The Northern people of Europe have, like the Canadians, a legitimate cause for concern. With Canada the complaint is legitimate too; much of the acid deposition it suffers does come from its industrially powerful neighbour, the USA. But of the acid reaching Scandinavia, 45 per cent comes from other European (especially Eastern European) combustion sources, including some from Norway, Sweden and Denmark themselves. The remaining 40 per cent comes not from land sources at all, but from microorganisms in the surface waters of the Atlantic Ocean and seas around Scandinavia.

The Scandinavians' goal of making the British fit sulphur removers to their power stations will produce only a 10 per cent reduction of acid deposition, at most. It is exceedingly unlikely that this will solve their problem. So why all the hype and the accusations that Britain is the dirty country of Europe? I think much of it is political. The voters in Norway and Sweden and Denmark are convinced by media disinformation that the British are to blame, and they put their governments under public pressure to pillory the victims and make them desist. The affair has an interesting feedback in the UK, where environmentalists almost welcome the accusation from abroad since it gives them a weapon in their battle with their own government.

Meantime there are grounds for believing that the other sulphur sources contributing to Scandinavia's acid rain are increasing. It will take time for Eastern Europe to curb its pollution and the emissions of sulphur gases by the ocean algae are probably rising too, as a result of the extra nutrients washing from farmers' fields down the rivers into the North Sea. This pollutant agricultural run-off, causing blooms of algae in the nutrient-enriched waters, may prove harder to curb than all the output of industry's chimneys. But the fact that farming, by increasing marine sulphur emissions, may be a cause of acid rain seems to be an unwelcome truth.

As Chapter 6 showed, the emission of dimethyl sulphide from the oceans has been known since the 1970s. It is not easy to measure DMS in the air, but the oxidation product methane sulphonic acid (MSA) is easily separated and measured from rain, mist, or the aerosol of acid droplets present in dry air. From the proportion of MSA to sulphuric acid in the total aerosol collected, the proportions of acid coming from the ocean and from combustion can be calculated. There is another independent way of distinguishing between sulphuric coming from algae and acid coming from combustion pollution. It happens that a subtle property of the atoms of sulphur, the ratio of the isotopes ^{32}S and ^{34}S, is different between natural and the manufactured sources. By measuring this ratio for the sulphur atoms of the collected sulphuric acid, it is possible to say how much came from combustion and how much from the ocean. A competent analytical chemist would not regard this as a difficult problem. You could easily be wondering: if this is so, did not the Swedish and Norwegian scientists make accurate scientific measurements early in the acid rain affair, some 15 years ago, and gather evidence that proved they were right – that, in this case, the input of acid deposition from combustion was much greater than that from the acids of oceanic origin? They could have done; they should have done. But in fact no such rational measurements were made. Surely then, even if the Scandinavians did not measure the acids of the rain, the British must have done so and put themselves in the clear?

They should have done. But this was no rational scientific investigation. It was tribal war. It has seemed throughout this affair as if no one on either side wanted to know about the real sources of the acid. The continuance of the battle provided a focus for environmental concern, as well as employment for air pollution experts carrying out routine research that could be filed, reported, published, and discussed at international conferences into the foreseeable future. At such conferences I often asked that the algal emissions be taken into account and that the composition of the acid falling on Scandinavia

CASE HISTORY Acid Indigestion

Acid rain is, at present levels, a comparatively minor affliction for Gaia, like the heartburn of acid indigestion. Nonetheless, a planetary physician would know that raised acidity is harmful to the well-being of living organisms, and while minor in itself may be an indicator of an underlying pathology. Human concern over the issue is often centred on political interests – such as the commercial value of forests, fisheries, ancient buildings, and tourism. The physician would want to look at the acids in the air more objectively, to assess their true sources in metabolic disturbance to the system.

So what are the acids of the rain? The table below lists the main acids in clean and polluted air. Of these, the strong acids deposited as rain or mist on the trees (see left) are nitric, sulphuric, and methane sulphuric acids. Nitric acid is present in clean air as a result of lightning (see p.119) and of the emission of nitric oxide from a wide range of biological sources, but it is too dilute in clean air to be of concern. In urban regions, cars and other transport and combustion sources generally emit nitrogen oxides that finish up as nitric acid and this does contribute to acid rain.

More important are the sulphur acids. The chimneys of coal-burning power stations emit large quantities of sulphur dioxide high in the air and this gas moves with the winds, slowly oxidizing as it goes to sulphuric acid. The algal life of the ocean is responsible for as much sulphur as industry. Both emit about 100 megatons a year. The principal algal emission is dimethyl sulphide or DMS (see Chapter 6), which oxidizes in the air fairly rapidly to form sulphuric and methane sulphonic acids. Methane sulphonic acid is uniquely oceanic in origin.

Acid rain is afflicting new areas, from China to Chile, a result not only of industry and the automobile, but also of forest and grassland burning. The planetary physician might suspect agriculture too: in European seas, emission of algal sulphur is particularly strong near the coasts where rivers dump the farm chemicals washed off the soil; the same may become true elsewhere. And raised sulphur output from ocean algae may have other consequences besides acid rain, if the link with climate (see Chapter 7) proves a reality.

Sources of the acids in clean and polluted air

Acid	Clean air	Polluted air
Carbonic	Natural CO_2	Natural plus combustion CO_2
Formic	Methane oxidation	Increased oxidation
Sulphuric	Ocean sulphur gases	Combustion
Nitric	Lightning, fires, etc	Combustion
Methane sulphonic	Ocean algae	–

be measured so as to indicate the probable sources. These attempts to introduce an "acid test" for acid rain failed.

Perhaps this was because the measurements of sulphur emissions from marine algae is difficult: it can be done, but the equipment must be devised by the scientists themselves; there is no ready-made commercial apparatus available. Or maybe it was just easier to ignore the tiresome subject in the hope that it would, if neglected long enough, simply go away. Even UK industries, who stood to lose profitability if they were singled out as prime culprits, were uninterested. If we thought that tales of acids coming from the oceans would stop the tribal war, we were whistling in the wind. Truth was the last commodity needed in the European battle over acid rain.

None of this would matter much, apart from the discomfiture of the British at having their country made the scapegoat of Europe, were it not that acid rain is harmful and should be stopped. The experience of Europe does not bode well for human responses to the problem as it spreads to developing countries in other regions of the world. If even friendly nations squabble over who is to blame, while ignoring the possible true sources of the damage, how will less-comfortable neighbours fare? On the larger scale, a great deal of energy and resources are wasted in the anger and the fighting of these tribal battles over what are, from a planetary viewpoint, insignificant affairs. The use of environmental discomforts to score points in national politics is a superannuated luxury. It is as irrelevant as conflicts among humans over religious and scientific fundamentalism. Now, at the start of the 21st century, there is little time left to remedy real problems affecting the Earth. The global environment encompasses us all regardless of race, creed, national pride, or political persuasion. I am glad to report that in the 1990s the nations bordering the North Sea at last saw sense and measurements of algal emissions were included in the European sulphur budget. Much of this came from Professor Peter Liss and his colleagues at the University of East Anglia.

The ozone war

Perhaps, after reading my views on acid rain, you may feel that I exaggerate or distort the truth. If so, there is another environmental battle, happily now almost solved, but which at its peak was just as combative and illustrated the need for a common sense approach, such as that of a physician. The battle I have in mind was over the right action to prevent the depletion of stratospheric ozone by halocarbons. Let me begin by telling you something of my personal experiences in what has been called the Ozone War.

The story begins in the late 1960s in Western Ireland in a holiday cottage looking out over the Atlantic. Most of the time the air was so clear you could see the islands 50 miles away; but occasionally it was very hazy. I had the idea to find out if the haze was natural or manmade by measuring its chlorofluorocarbon content, CFCs being unequivocally of industrial origin. Using a home-made gas chromatograph equipped with an electron capture detector I first tested the sparkling clear air, and was surprised to find one of the CFC gases present at about 50 parts per trillion. A few days later came the haze; it contained three times as much CFCs as the clear air. So it *was* man-made – as we later found, it was smog from southern Europe carrying the exhaust fumes of millions of cars of holiday makers.

I continued to wonder about those 50 ppt of CFCs in the *clean* air. Had they drifted across the Atlantic from America? Or were the chlorofluorocarbons accumulating in the Earth's atmosphere without any means for their removal? To find out, I would have to measure the CFCs across the world, over the oceans, and in the Southern Hemisphere. I applied for funding but was turned down by the scientific review committees. This was a discouragement, but not a deterrent. I managed to go on the expedition anyway, due to the generosity of my first wife, Helen Mary, and of the civil servants of NERC who let me travel to Antarctica and back on one of their ships, the RV *Shackleton*. Thus I was able to use my home-made apparatus to measure CFCs across the hemispheres.

The discovery of the global distribution of the chlorofluorocarbons can be credited to small-scale, old-style science: the apparatus was so simple that I was able to make it in a few days, and the total cost of research was a few hundred pounds. The important point is that such an expedition would never have been mounted at all by the armies of big science, who nevertheless within a few years of the discovery were spending tens of millions on the same topic.

The report that CFCs were accumulating in the atmosphere, almost without loss, stimulated Sherry Rowland and Mario Molina to write their historic paper in *Nature*, warning that the breakdown of the CFCs in the stratosphere would release chlorine, and that this could catalyze the destruction of stratospheric ozone and hence increase the flux of ultraviolet radiation at the Earth's surface. I was very sceptical at the time. Not of the validity of their hypothesis, but of its importance. The 50 ppt of F11, and the 80 ppt of F12 then in the air (in the mid-1970s) seemed to me to be things to keep a watch on, no more than that.

But once the popular media realized that yet another group of industrial chemicals, no matter how harmless and useful, might be the indirect cause of cancer – because ozone depletion could increase our exposure to ultraviolet, which is a source of skin cancer – there was an outburst, an explosion, of hype. Newspapers carried banner headlines saying: "These chemicals will destroy all life on Earth". Funds flowed as never before. Something near to $1000 million must by now have been spent by big science on stratospheric research connected with the CFC ozone affair. This time the chemical industry was amazingly cool, actively supporting the work of the scientists – including those whose findings were in favour of a cessation of production. The CFC battle was fought, not between industry and environmentalists, but between rival scientists jealous of their right to defend that sacred place in the sky.

As a consequence of the Ozone War, big science has been able to improve our understanding of the complexities of the atmosphere. But it is a scandal that the vast sums spent on computer models of the stratosphere and on satellite, balloon, and aircraft measurements failed to predict or find the ozone hole. Instead, it was discovered by three British observers, Farman, Gardiner and Shanklin, using an old-fashioned and inexpensive instrument on an expedition in the Antarctic, rather as was the presence of CFCs themselves. In fact, the computer modellers programmed the instruments aboard the satellite observing atmospheric ozone to reject any new data that was substantially different from the model. The instruments saw the hole; the scientists discounted it. The Ozone War is littered with stories of this kind.

The public have been misled into thinking that CFC ozone depletion is the most serious environmental concern. While it was, and still is, extremely interesting to scientists, it is only a part of the environmental threat that looms ahead. But what about the ozone hole? Isn't this a serious menace to life on Earth? Won't it spread? I accept that ozone depletion could become a problem; but the dangers from ultraviolet are more to people, especially pale-skinned people, than they are to the planet. Moreover, the risks are still unclear. Only a few years ago, in Washington, I was told that the instruments measuring ultraviolet radiation at the Earth's surface must be faulty because they all showed a decrease in ultraviolet and therefore less danger of skin cancer. I should not like to be a passenger aboard an aircraft with these scientists as pilots, if that was how they reacted to messages from their instruments. Subsequently, a paper has been published in *Science* confirming that the instruments were right: ultraviolet radiation has *decreased* in intensity continuously throughout the last ten years of measurements in the continental USA. Scientific truth, as always, is complicated and hedged with qualifications. The UV measurements were mostly made at airports and therefore subject to the local pollution of the neaby city. We do not know how the UV levels have changed in unpolluted air.

In the almost 30 years since the prevalence of CFCs in the air was discovered, they have increased in abundance by 500 per cent. Since the ban on their emission at the Montreal conference of 1989 they are beginning to decline in abundance. I now have no doubt that it was right to stop the release of these gases. Not because of their ozone depleting tendency alone, but because stopping the emission of CFCs is the one positive thing we have done about the menacing greenhouse gas accumulation. CFCs are potent greenhouse gases.

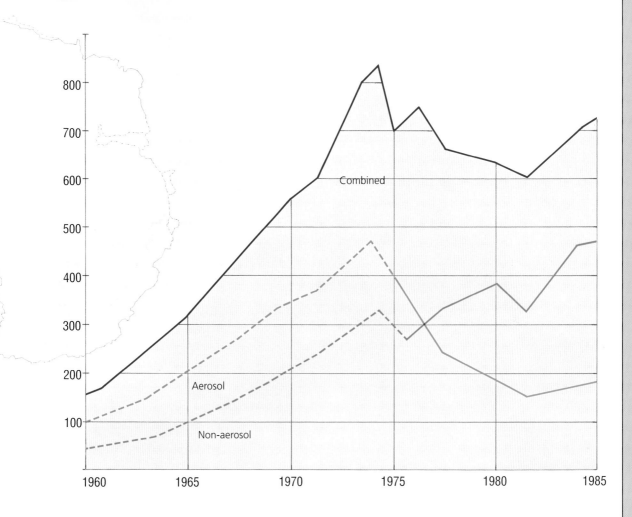

CASE HISTORY Anozonemia

For Gaia, ozone depletion and the ozone hole over the Antarctic are not the most serious threats. But to a planetary physician the presence of chloroflourocarbons in the Earth's atmosphere would be an unequivocal sign of disturbance to the planetary organism: these chemicals are not found in nature. CFCs are made by humans for three principal purposes: as aerosol propellant gases, in foam and in refrigeration. Public concern over ozone depletion and the risks of increased cancer rates due to ultraviolet exposure has led to strict controls on CFC outputs. The graphs show production of CFC-11 and CFC-12, as aerosols and non-aerosols and as the combined total. CFCs were still rising in 1985. Now the CFC abundance is at last falling but it will take decades to fall to safe levels.

The physician would be concerned about their effects as greenhouse gases as much as their damage to the ozone layer. Even with declining CFCs the ozone hole might still be with us, because it is also a consequence of rising methane – and methane is mainly the product of farming and forestry. Though not often mentioned a pollutant, it is probably the most dangerous substance that we are injecting into the atmosphere: a key agent in the ozone hole and, much more seriously, a greenhouse gas that has the potential to overtake carbon dioxide in significance. The production of CFCs is now banned altogether. Carbon dioxide, too, can be cut back, if we have the will. But to stop the excessive release of methane gas from rice paddies and from cattle is all but impossible.

I hope that this account of the Ozone War has shown that popular campaigners, while they often express hostility to science – and may be right to be sceptical of big science in particular – also reflect the prejudices of the scientific community, which itself is divided and uncertain. Scientists are human and for the most part concerned with careers, pensions, security, and all the needs of modern life. It is also not easy to be responsible where there is no accountability.

Yet optimism is justified by the extraordinary level of international agreement over the banning of CFCs and related chemicals. The response of the chemical industry too is developing ahead of time. I think that we are a responsible and responsive species. Our problem is just that our signals are confused and we are slow to act.

The warming Earth

We hear much these days about the greenhouse effect: of how the Earth may be heating up and the sea levels rising as the polar ice melts, and how great storms such as those of the past few years, might in some way be a consequence.

What concerns us, of course, is not the *existence* of the greenhouse – as Chapter 7 showed, this is really a good thing, a part of Gaia's temperature regulation – but the way that new panes of heat absorber are being added to it. Many of the pollutant gases we release to the atmosphere are powerful absorbers of infrared radiation, and by retaining more heat can increase the warmth of the Earth. If we continue to add these gases to the air, in time the Earth may become too hot for our comfort.

The facts about the principal greenhouse gases – their present abundance, likely increase from human sources, and percentage contribution to any warming effect – are tabled in the case history (p. 169). If we know all this, isn't it easy to calculate the additional heating due to our actions, and predict just how hot it will be in a few years time?

Not so. We would first have to understand many climate feedback effects – of water vapour and clouds, of ice and snow cover, of the oceans, and most of all of the biota – effects of which we are still profoundly ignorant. And then we would have to be able to model all these interactions, along with the greenhouse gases and their effects. The affair is so complex that accurate prediction is impossible.

One problem, as always, is the fragmentation of science. Meteorologists, for instance, have built super computer models of the atmosphere, with the spherical shell of the air divided in three dimensions into parcels a few kilometres in size. These models take into account all of the atmospheric gases and their greenhouse effects and also the effects of land and sea surface temperatures, ice cover, and the effects on winds of mountain ranges. But the clouds, though included in the models, are usually treated as more or less fixed entities unable to vary much with the changes of climate. When these models are used to predict the rise in temperature due to, say, a doubling of the greenhouse gases (something that we already know could happen in the next century) they predict a global temperature rise of between 3 and 6°C.

None of these models yet take account of the cloud–algae connection proposed by Charlson, Andreae, Warren and me (see Chapters 6 and 7). While we suspect that there may be an active link between global climate regulation and the algal production of DMS, we have no clear idea of its nature. All we know is that during the last glaciation the emissions of DMS by ocean algae appear to have been two- to four-times greater than now. The capacity of this newly discovered potential climate control mechanism is also unknown. It could be insignificant, or it could have a cooling effect as great as the potential of the greenhouse to warm. Until a great deal more information about the oceans, the clouds, and the organisms living at the sea surface is available, it is not possible to create accurate and realistic enough models to predict the combined effect of clouds, algae, and the rise of greenhouse gases.

The eminent geochemist Wally Broecker warned in *Nature* in 1987 that there could be "unpleasant surprises in the greenhouse" due to possible unexpected geophysical positive feedbacks that would render ordinary prediction impossible. He mentioned the rapid clearance of ice cover at the poles, and the potential instability of the great masses of cold and warm water in the oceans. Anything large that happened to these systems might provide a nasty surprise. Broecker's nasty surprises could – if the amplifying power of the rapidly responding system were coupled into the geophysical process – be intensified to become catastrophes.

When I first wrote in 1990 I said: "When governments ask scientists what will happen as the greenhouse gases rise, and how urgently we should act, it is hard for them to know what to say. As we have seen, the affair is very complex. And though their theories tell them it should grow warmer, when they look for confirmation in the temperature record, there seems to be little evidence that the increase of greenhouse gases so far, in the last 200 years, has done anything." Now in the year 2000 we have ample evidence of global warming. The Northern Hemisphere appears to have risen by 1°C in the past 200 years reversing what was for the previous 1000 years a cooling trend. In addition we can say, first and most seriously, the greenhouse gases are inexorably increasing and in the long run, even if they never rose above the present levels, would change the world irreversibly. They are like one of those nasty poisons that acts only months or years after taking it. To understand what we have already done we need only look back to the last glaciation. Apart from a time lag of a few hundred years, temperature and carbon dioxide go closely together. At the depth of the ice age the CO_2 was

180 parts per million (ppm) and the global mean temperature 5°C lower than now. At the start of the present warm period the CO_2 was 280 ppm, so that the addition of 100 ppm of CO_2 corresponds to a rise of about two degrees. We have added 88 ppm of carbon dioxide and we have also doubled the methane and added enough CFCs to increase the gaseous greenhouse by an amount equal to about 40 ppm of CO_2. Taken together we have changed the atmosphere more already than took place between the last glaciation and now. The consequences, a rise of temperature comparable to that between the glaciation and now, are probable.

The second thing to keep in mind is that, with Gaia present, the system might react to amplify heating and aggravate global warming. At the start of the Daisyworld model, remember, the growth and spread of dark daisies were tightly coupled to the geophysics of the planetary radiation balance; temperature and dark daisy growth rose rapidly together with positive feedback. In the fevered state of the interglacial this is the expectation of Gaia's intervention now.

Historical evidence taken from core samples of the Greenland ice cap reveal that the rate of rise of temperature at the transition from the last glaciation until now was so rapid as to require something more than a mere geophysical explanation. The transition had the hallmarks of a Gaia system response – of geophysiological surprise. The surprise that Gaia may produce now, in response to the added greenhouse gases, is what we have to fear.

The third thing to remember is that the time constant for the CO_2 cycle is very long. Any change we make will not go away rapidly.

The will to live in harmony

These are some of the consequences of the overabundance of people on this planet. It is not the number alone that matters. It is what people do, and what their crops, livestock, pets, and dependent species do. The metaphor of planetary medicine may help us to understand our place in the etiology of the planet's ailments. But a metaphor can never be accurate. If it were it would be a replica or a recipe. For example, with the planetary disease, *Disseminated Primatemia*, the superabundance of humans, it is the disease agents (the people) that are sentient and the host (the planet), a lowly organism – the reverse of the situation with human disease. But to reach the desired states of symbiosis and homeostasis, whether individual humans with their microbial diseases, or planets infested with people, it is the sentient partners, the people, who need the will to live with their partners in symbiotic harmony. Whether humans are the host or the parasite seems to matter less.

Gaia's fever

As the graph above shows, we know more about Gaia's past temperature than about the future. Since pre-industrial times we have increased the greenhouse gas content of the atmosphere by more than the natural increase that occurred at the end of the last Ice Age. While we cannot predict the outcome with any certainty, a comparable rise in temperature – of 5°C – is likely. It could be a sudden jump to a hot and inhospitable state.

CASE HISTORY The Fevered Planet

Organisms under stress are vulnerable to infection by microorganisms that normally would be present only at tolerable numbers. A planetary physician would view the greenhouse effect as just such a pathology afflicting Gaia – one where the stress of interglacial warming had allowed rapid multiplication of humans and their civilizations, with aggravating effect. Temperature records show clearly that the Earth's temperature is high, though not as yet higher than in past episodes of interglacial fever from which the planet has recovered. More worrying are the signs of damage to the forest ecosystems that normally would provide the planet's natural cooling mechanisms, and the biochemical disorder signalled by the rise of greenhouse gases, including the specific human toxins, the CFCs.

The chart below details the human sources and the rising abundance of the principal greenhouse gases, together with their effectiveness as infrared absorbers, and their percentage contribution to any warming effect. Even with all this information, however, it is not easy for our planetary physician to predict the course of the fever. Take the greenhouse gases themselves. As you can see, carbon dioxide is by far the most abundant; released as a by-product of *all* combustion processes, CO_2 pollution is the price human civilization pays for its need for energy. But by far the most effective of the greenhouse gases, molecule for molecule, are the CFCs. These gases are not only a danger to the atmosphere through their potential to destroy stratospheric ozone – they are also very potent infrared absorbers, about 10,000 times more so than carbon dioxide. Moreover, they absorb infrared at wavelengths between 8 and 13 thousandths of a millimetre, a part of the spectrum where the atmosphere otherwise is transparent. So, despite their low abundance, the CFCs are a serious and growing element in global warming.

Another potent greenhouse gas, whose action is complex and difficult to predict, is methane. It may be that in the next century methane will be the most significant of the greenhouse gases. During the 1990s we have grown aware that warming the oceans decreases the areas covered by dense algal blooms. Algae like cool waters below 12°C because warmth stratifies the ocean and denies the algae the nutrients in the cooler water below. In a similar way land plants prefer temperatures cooler than the growth optimum because warmth causes the faster evaporation of rain and so denies plants the water they need. Lee Kump and I in a *Nature* paper in 1996 argued that together these two effects act to amplify global warming. Gaian systems do not always ameliorate adverse conditions, as in a fever the normal response is sometimes reversed.

Greenhouse Gases

Gas	Pre-ind Conctns ppm	Present Conctns ppm	Possible Conctns by 2030	Warming Contrib. %	Warming Effectiveness	Human Sources
Carbon dioxide	280	368	400–500	49	1	Combustion of fossil fuels – coal, oil, and gas. Deforestation and changing land use. Biomass burning
Methane	0.7	1.7	1.85–3.30	18	25	Wetland agriculture. Enteric fermentation in cattle and termites. Leakages from gas and oil exploitation. Biomass burning
CFCs	–	CFC–11: 0.0002	0.0005–0.002	14	CFC–12: 10,000	Used in refrigeration, air conditioning, plastic foam, and as propellant, solvent, sterilant
	–	CFC–12: 0.0004	0.0009–0.0035			
Nitrous oxide	0.28	0.31	0.35–0.45	6	150	Nitrogen-based fertilizers. Fossil fuel combustion. Biomass burning

Feedback loops

Here are some other reasons why predicting the effect of greenhouse gases on Gaia's temperature is not at all simple.

Water vapour and clouds Water vapour is a powerful greenhouse gas but the amount of it in the air depends on the temperature. If it is cold it is removed from the air as rain or snow. If it is hot much more water vapour can exist as a gas in the air. This complication can be dealt with but water vapour also condenses as clouds, mists, and fogs, and at all levels from the surface to the high stratosphere. Clouds near the surface tend to reflect light back to space, whereas those high up in the stratosphere reflect heat back to the Earth. In general, but not always, high clouds heat and low clouds cool. Ice formation at high levels adds yet another complication: the distribution of ice and water in clouds can sufficiently affect their radioactive properties to halve the predicted warming by greenhouse gases.

The oceans If clouds complicate our predictions, so do the oceans. The ability of the oceans to store heat is 1000 times that of the air. The extent of greenhouse gas heating will greatly depend upon how much of the added heat is distributed in the oceans. If all of the predicted rise of surface temperature were mixed in with the oceans uniformly, there would be no perceptible rise in surface atmospheric temperature during the next century or so.

The polar ice Snow is brilliant in its whiteness and reflects heat in the form of sunlight back to space. Snows tends to sustain itself by keeping cold. Increasing warmth should in a simple world melt the snow, leaving dark ground, that would then absorb sunlight and with positive feedback hasten the melting. In the real world there are mountain regions, like Greenland and the Antarctic plateau, covered with thousands of metres of solid ice. These regions are so cold that a few degrees of global warmth is not enough to melt them but, more important, increasing warmth brings more water vapour and more snow. Ice can give rise to negative as well as positive feedback on global warming.

The organisms The involvement of living organisms in albedo, and especially in regulating cloud cover, whether through evapotranspiration over the forests, or through DMS production in the algal fields of the sea, is a major complication in any prediction of climatic change. Living organisms are involved intimately too in the removal of CO_2 from the air, and methane production. All the great ecosystems – the photosynthesizers, the consumers, and the methanogens – have interacted with climate throughout Gaia's long history.

Obviously, the prognosis for the fevered planet is uncertain. What is certain, however, is that we have already administered enough poisons to make the patient sick, and Gaia's temperature is bound to rise, in the short term. We must not expect the Earth system to bring about a natural cure; more likely the Earth, sickened by what we have done, will grow more fevered still.

The runaway greenhouse

In the fevered interglacial, Gaia's ability to combat infection may be limited. A system failure could lead to rapid heating, as sudden as that which occurred at the end of the Ice Age. The system is already in positive feedback and further heating could create a geophysiological surprise. As warming melts the white poles, the dark ground and ocean are exposed to absorb more heat, albedo is lowered, and the tundra melts, releasing more methane, which further hastens the warming. As the ocean surface waters warm, algal growth is less so that the pump down of CO_2 and the production of DMS lessens. All of these effects are positive feedback on warming.

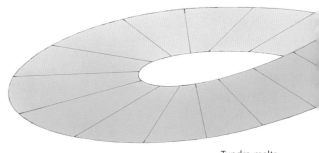

Further
temperature
increase

Tundra melts
Increased methane

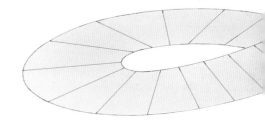

Icecaps
start to
melt

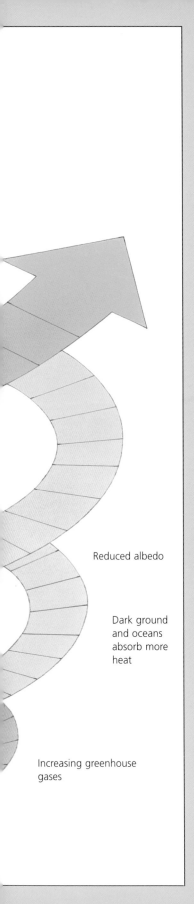

Reduced albedo

Dark ground
and oceans
absorb more
heat

Increasing greenhouse
gases

Conclusion

Lewis Thomas, in his book *The Youngest Science,* persuasively portrays the life of a physician in the days before 1940, prior to the time scientific medicine came of age. He describes how he travelled as a child with his father, a general practitioner, on his rounds. How the black bag his father carried contained only three drugs of proven therapeutic value, morphine, digitalis, and adrenaline and how only the first two of these were ever used. There were no cures then like antibiotics now. The only way a physician could favourably influence the course of the battle with a disease agent was to improve the patient's morale. Important among his skills was the act of prescription. On a special pad he always carried he would write out in Latin a list of ingredients for a nostrum. Always the patient or the relatives wanted a bottle of medicine, believing in a cure. Some of these nostrums were fragrant or aromatic, some, flavoured with asafoetida, malodorous. They had no intrinsic therapeutic potency; their value came from the almost religious ambience of the act of prescription. Like the bread and wine, they were dispensed in Latin and felt rather than understood.

Now, in these secular reductionist times, we seek to explain the mysteries of prescription by saying that the bodily defence, the immune system, is strengthened by feeling good, by the sense of contentment that comes from the sacred transaction between patient and physician. I don't know if any statistical studies have ever been made to determine whether or not old-fashioned medicine worked in this way. I do know that the good physician made living, sickness, and dying a great deal easier, even though equipped with little more than opium alkaloids for pain and to ease the process of dying, digitalis for broken hearts, and laxatives.

The main thing to keep in mind is that the practice of planetary medicine is like human medicine before the scientific revolution that began in the 1940s. We are in a state of profound scientific ignorance about the Earth and what science we do have is as yet hardly applicable for the solution of our environmental problems. For example, we know that the gases of the atmosphere have a

Human impact

With the remote eyes of satellites we can now look down and see clearly our own impact on the Earth. The Landsat image here is of the Ganges Delta, Bangladesh, at the end of the dry season. The blue tone at the river mouth is due to sediment from erosion by human forest clearance upstream. The red areas are dry crops, those dotted with red, rice paddies (great generators of the greenhouse gas methane). Only the small greenish areas are "wild".

profound influence on climate. This is important knowledge, but it is far from enough to enable us to propose a global system for climate management. Keeping a healthy planet with an equable climate is as difficult for us as was keeping a healthy body free of disease for our forebears. Microbiology was well established in the late 19th century, but it could do little or nothing to abate the scourges of tuberculosis, diphtheria, and virulent streptococcal infections. It took over fifty years before antibiotics and immunization techniques effectively eliminated these diseases.

The analytical scientific approach helps us to diagnose disease in the present embodiment of Gaia, but prescribing remedies is quite beyond our skill and dangerous to attempt in ignorance. This does not mean that nothing can be done; only that we should regard "clever" quick technological fixes as no better than patent medicines.

In the last century, the environmental problems were just as serious as our own today. In the mid-19th century there were epidemics of the water-borne diseases, cholera and typhoid, that caused the death of a third of the inhabitants of some cities in a few months. Science was not then organized as a powerful lobby and was prepared to admit that it did not know the cause of the diseases. Physicians at the sharp end of this battle suspected from the epidemiology that infection was water-borne or came from the bad odours of the primitive sewerage systems then in use.

Our practical forebears did not pour funds into the infant science of microbiology and wait until it proved that cholera and typhoid were water-borne bacterial infections. They acted promptly and empirically by installing clean water supplies and efficient sewerage collection and disposal plants. Engineering was in those days a proud profession and triumphantly displayed its self-confidence in those amazing gothic pumping stations that are now a place of pilgrimage for students of industrial architecture.

The scientific background for the 19th-century physician was physiology. We can trace the recognition of the idea of self-regulation in a physiological sense to the French scientist, Claude Bernard, who sowed the seeds that led that great American physiologist, Walter Cannon, to the concepts of homeostasis. It was this kind of science that encouraged the good physician to intuit that the body was wonderful in its capacity to take care of itself. Often, lacking any rational therapies, it was best to let Nature take its course, with a little practical help in the way of setting broken bones and other acts needed to repair any damage done.

Present-day good physicians, although conversant with biochemistry and microbiology, still recognize that living healthily is what really matters. They avoid the ostentatious prescription of powerful remedies, such as antibiotics. They prefer to keep them as a remedy of last resort and know that their unwise and too generous use can lead to a state worse than that before their discovery.

The principal causes of death – heart disease, strokes, and cancer – are still effectively incurable. But their age of onset can be postponed by living well. More and more it seems that the real role of the physician is not to cure disease but to prevent it happening, so that we can enjoy in full health our allotted span. In the same way, the role of a good planetary physician should be to persuade you of the advantages of living well with Gaia. To make you feel for Gaia, or for the Earth, as you would about your own body. At this stage of our knowledge of the Earth, prescriptions are of no more and no less value than were the nostrums of the old family doctor who made house calls on the sick.

Living with Gaia

If we accept that we humans have a finite individual lifespan, and that no one can ever be immortal, then maybe we should keep in mind the thought that our species also has a limit for its span on Earth. Instead, in our optimism we imagine that if we could manage ourselves and the Earth well enough we could, somehow, find ways of coping with a doubling of lifespan, or a doubling of population. We assume that the extra stress we should then place on the Earth's ecosystems could be prevented or alleviated by good stewardship or planetary management.

I think that this is the greatest of our errors. Consider how the well-intentioned application of the principles of human welfare and freedom that moved us all in the second half of the twentieth century has failed our bright expectations. Cruel tyrannies now reign in much of what has been labelled the Developing World. In spite of modern medicine, in many places the quality and the length of life diminishes as the land dies under the weight of sacred cows and insupportable numbers of people.

Consider also yourself. You might suffer the misfortune of an accident that damaged your kidneys. Not fatally, but enough to cause those wonderful intelligent filters to fail in their task of regulating the electrolytes, the salts of your blood. You can survive, even live a normal life, but only by always taking care to monitor your intakes of salt and water. A burden of this kind powerfully reinforces the wonder at how well our body manages itself when we are healthy. With disabled kidneys you would have to be the steward, the manager of your body. A permanent employment, not difficult, but life would no longer be carefree. An invitation to stay with friends becomes a problem of salt balance, as would hard physical work, or a brisk walk, on a hot day.

But this example is a disablement of one system only. If several bodily systems were disabled simultaneously then you really would have little chance to do anything but consciously regulate your bodily functions. This is the kind of burden, slavery, I have in mind when I say there is no worse fate for humans than to so disable the

Earth that to survive they must take on the task of running the planet. Just think of the task of managing even a developed nation, so that the emission of carbon dioxide from burning fuels and agriculture was balanced by the uptake from planted trees. A task that would require the meeting and matching of the conflicting interests of the individuals and groups that make up human society, the resisting of the powerful selfish pressures of their lobbies, and at the same time coping with the haphazard changes of the political, economic, and actual climate. That would be just the start of it, for then there would be the same and other problems involving the inputs and outputs of your nation with those of the numerous other national and tribal states of the world.

A planetary physician can only prescribe for your relationship with the Earth that kind of love and benign neglect that characterizes the relationship of good parents toward their children. There are no nostrums or simple remedies for the ills of the Earth.

This does not mean that there is nothing that you or your society can do about the health of the Earth. A good parent does try to provide an environment that is not damaging to their children and allows them to gain the strength to heal themselves. There are many simple things that each of us can do to live better with Gaia. We cannot manage the Earth, but we can usefully regulate our own lives, and our human institutions. I find it helpful as a start to keep an image of the three deadly Cs in mind: Cars, Cattle, and Chainsaws. We need not be fanatical and ask for them to be banned; it wouldn't work. But we can remember the physiological truth that the poison is in the dose, and be moderate in our use of these and other dangers to the health of Gaia.

The advantages of moderation in the use of cars and chainsaws are immediately self-evident. The damage wrought by excessive cattle farming, though less obvious, is equally severe: to produce food as beef or dairy products requires twenty times as much land as its vegetable equivalent. I do not propose that we all try to become vegetarian. Better first to think about Africa. We know that there is frequently famine there, yet few seem to realize that much of this distress comes as a direct result of land damage by primitive cattle farming. The human and natural ecosystems of that unhappy continent may soon disintegrate. In Africa, it is not overpopulation with people that is the problem but overpopulation with livestock.

There are other ways of living better with the Earth. Most of them are personal and I do not see this book as the place to list them. You would not expect to find advice on healthy gourmet cooking, or maps of places for walking to keep fit, in a textbook on human medicine. These are for you to find for yourself, using your own judgement on how best to moderate your demands on the Earth, and yet enjoy life. In the same way, you do not need me to tell you all the positive actions you could take – from planting trees wherever you can, to helping clean up the environment where you live and work. There is no shortage of advice on how humans can as individuals live healthily with the Earth.

There is plenty of advice around, too, on how we could collectively, as governments and other institutions, act to solve the "environment crisis". Some of this advice is in principle good. We should, indeed, stop clearing the forest, reduce industrial and other pollutants, develop energy efficient solutions, cut back on fossil-fuel burning, seek less-damaging agricultural techniques, and try voluntarily to curb our numbers and consumption. But in practice, even if we find the will to act, the trouble with much of this advice is that, like some invasive medical approaches, it may do more harm than the disease. Our thinking is still deeply human-centred, based on short-term self-centred advantage, an overestimate of our powers, and a profound ignorance about the Earth.

I would particularly warn against imprudent planetary medication or surgery. There are, for example, proposals to "cure" the effect of greenhouse gas poisoning by applying medication to the oceans, to stimulate the algae there so that they remove the excess carbon dioxide from the air. By irrigating the oceans with iron chloride solution, dispensed from supertankers, we could, say the experts who dreamed up this idea, fertilize the algal blooms and remove enough carbon dioxide from the air to allow us to continue burning fossil fuels without restraint. By a fluke, this scheme might in the short term achieve its primary intention, of reducing carbon dioxide in the air. But it would still be foolish – as unwise an act as taking thyroid hormone to increase one's metabolic rate, so that a fancy for sugar, cakes and hamburgers could be indulged without the penalty of obesity. Both prescriptions – of iron chloride for the planet, or of thyroid hormone for the fat person – fail completely to recognize that the patients, whether Gaia or a human being, are self-regulating living systems. To attempt control from outside by increasing or decreasing one feedback loop only in these multiple feedback systems is rarely successful, and carries with it the risk of dangerous and unpredictable instability.

I have tried to show in this book that the carbon dioxide and climate balance for the present state of Gaia is unstable and that carbon dioxide is oscillating at a level only just above the lower limit for the growth of plants. For a system near the limits of stable regulation the effects of adding or subtracting carbon dioxide are unpredictable. The only wise course would be to cut back the emissions of greenhouse gases.

What do we know of the Earth?

I have made a strong statement rejecting the idea of planetary management, or stewardship, in so far as it implies taking charge of the Earth. I propose instead that we learn to live with the Earth as a part of it; by managing ourselves, and by humbly taking and giving the gifts that sustain all of us who live on this planet.

Some of you may regard my proposal as irresponsible. As the only organized intelligence, surely we have the duty, if not the right, to take charge of the Earth and govern it responsibly? Maybe so, but how can we manage it if we do not know what it is? First we must ask, what is the Earth? This may seem a trivial question – everyone knows what the Earth is – but unfortunately there seems to be no common view. Because we take the Earth for granted we tend to act like bacteria, never noticing the consequences of our unchecked growth. Even scientists differ about what the Earth is, although now in the year 2000 more scientists see the Earth as a whole. But many, even if they give lip service to either Gaia or coevolution, still act as if the Earth were a ball of white hot, partially melted rock with just a cool crust moistened by the oceans. On the surface they see life as a thinly spread layer or organisms that have simply adapted to the material conditions of the planet. With such a view go metaphors like "the Space Ship Earth". As if humans were the crew and the passengers of a rocky ship forever travelling an inner circle around the Sun. As if the 3.8 billion years that life has existed on Earth were just a prelude to the evolution of humans and to serve as their life-support system when they chanced to come aboard. Seen this way, obviously the Earth might appear fragile. Those who so see it must wonder how it has survived so long.

This is the conventional wisdom about the Earth, and is still taught in most schools and universities. It is almost certainly wrong and has arisen as an accidental consequence of the fragmentation of science, into a growing collection of independent scientific speciali-ties. There are a few geographers and Earth system scientists who teach or research the Earth as a whole. But the majority of practising Earth and life scientists are specialists and even though they know that the conventional wisdom is wrong, they continue to take their views of the Earth from it. If we want to know about life, the universe, and the Earth, we read about them in the *New Scientist* or the *Scientific American*. Back in the laboratory, scientists continue in their own speciality without concern for either the general wisdom or the intricate details of the specialities of their close colleagues.

Of course, no single approach can lead to a complete under-standing of the Earth, all are needed. We need the reductionist model of the Earth to understand details at the molecular level. A key example is the chemistry of the stratosphere. It was only through the application of classical atmospheric chemistry and physics that Rowland and Molina first made known the threat to ozone from the CFCs. From biogeochemistry there came, through the work of G E Hutchinson, the recognition of the role of micro-organisms in the soil and the oceans as the source of methane and nitrous oxide. From geophysiology came the recognition that atmospheric gases, such as carbon dioxide, methane and dimethyl sulphide, may be part of a physiological climate regulation.

We are at a time when many scientists as professionals seem to have lost sight of the Earth as a planet in the intricacies of detail. As a result they may often be more concerned about specific dangers to people, than about the looming threats to the planetary environment. The foremost personal and public fear is that of cancer. Consequently, any environmental chemical or radiation thought to cause cancer is given attention out of all proportion to the real risk it poses. Nuclear power, ozone depletion, and toxic chemicals such as dioxin and PCBs are regarded as the most serious of environmental hazards because of this fear (and also because nuclear radiation and halocarbons are so easy to measure). I think that the potential hazards of the gaseous greenhouse and land abuse have, until recently, been ignored because they perturb the planet, not individual people, and because they are much more difficult to quantify.

The art and science of model building has matured in the ten years since this book was first written. Global climate models now exist in the world's climate centres that take account of greenhouse gases, and include land-based plants and ocean algae. Clouds are still a difficult problem and I am not yet convinced that more than a few models are tight coupled, geophysiological and include the oceans properly, but we have come a long way towards predicting future climates. We have made progress but we are still far from understanding the Earth as a system.

So how can we govern it if science is still unable to tell us what it is? Should we wait for the deliberations of the plenary session of the all-science interdisciplinary congress? Or should we listen to thoughtful environmentalists such as Jonathon Porritt, who ask: can we afford to wait for scientific certainty before taking the obvious sensible action on environmental affairs?

The need for empirical approaches
Consider for example the accumulation of chlorofluorocarbons in the atmosphere. No one doubts that CFCs had, before they were banned, reached a level that is damaging and we are all agreed that it was right to ban their emission. Sensible Greens are puzzled about why, if this is so, we continue to spend billions of scarce funds on stratospheric and ozone depletion research, when the problem is in effect solved. We know the poison, all that needs doing is to stop imbibing it. If we were serious, say the Greens, we should be considering the general problem, how do we refrigerate and air condition without letting loose CFCs to the atmosphere? How do we dispose of the large stocks of CFCs in storage and in the refrigerators now in use? Compared with the excitement and glamour of research in the upper atmosphere, of exquisite physical chemistry experiments, or of elegant computer models, CFC disposal is a problem for mere

engineers, the rude mechanicals of science. Proof of the lack of this kind of simple engineering is the fact that a substantial proportion, possibly more than half of the CFCs that entered the atmosphere, came from leaking air conditioners in American cars. The Greens are right. Engineering research should be at the top of our lists for action instead of at the bottom as it is now.

We don't need to burn a billion candles to specialist scientific research. It only pollutes the air and in the end may only confirm that it is too late to postpone our doom. Let us spend our cash now with the planet, not people, in mind. Otherwise we are like a farmer who would mortgage his land to pay for his children's education in a nearby city, not realizing that he has ensured their separation from the land and unwittingly denied them their inheritance.

But the Green movement itself can be a potent force preventing environmental reform. Some sections of it have anachronistic views of industry, regarding it as irredeemably harmful and polluting. Among them are those purists who see all industry as inherently bad, and driven only by profiteering and power. Others would return to a romantic but impractical rural existence. This kind of Green nonsense is encouraged by the tendency of talented writers and dramatists to cast their villains as owners, or employees, of the nuclear or chemical industries. They go on to make these industries the stage equivalent of the desecrated graveyard of a Victorian melodrama, places of quintessential evil.

This flood of righteous wrath ignores the certainty that if we gave up our industrial civilization only a few of us would survive. More seriously, it distracts attention from the real and urgent need for industry to reform and become non-polluting, low-consuming and benign, the basis of a new civilization in harmony with Gaia.

We could go along with the Greens, and still retain the long-term guidance of science, if we could first delineate the near certainties and then act empirically in an engineering way. We now know, for example, from the record of the gases trapped in the layer of Antarctic ice, just how the atmosphere has changed during the past 200,000 years. We know that the carbon dioxide 15,000 years ago in the depths of the Ice Age was about 180 ppm and that at all times there has been a close correlation between carbon dioxide and temperature. The rise in temperature and the rise in carbon dioxide over a period of only a few hundred years at the end of the Ice Age was 3°C and 100 parts per million respectively. In the past two hundred years we have increased the atmospheric carbon dioxide by 80 parts per million, and if we take the effect of the CFCs into account, we have already effectively increased the greenhouse effect by as much as happened between the Ice Age and the interglacial that followed. We suspect that the mean global temperature will rise, even if we stopped emissions now. The uncertainty is less about the rise in temperature than about how fast it will take place.

Knowing this much, can we emulate our forebears and develop an empirical approach to planetary problems? We can, and there are many practical things to be done. Fossil-fuel burning power stations are among the principal atmospheric sources of CO_2, SO_2, and nitrogen oxides. They are at best only 40 per cent efficient and waste the rest of the energy of the fuel in those ugly cooling towers that advertise their presence. There is no reason on scientific and engineering grounds why coal should not be burned so that most of the energy is made usefully available, and so that the pollutant emissions are gathered and either used profitably or stored where it will do no harm. Pilot plants in which coal can be converted to hydrogen and carbon dioxide already exist. In these plants, the hydrogen would be the fuel of gas turbine power stations and the carbon dioxide would be collected and disposed of underground or in the ocean. At present, although more efficient in energy terms, these alternative power sources are not competitive in cost with conventional power stations. But no sensible engineer would expect pilot plants to compare in efficiency with the evolutionary product of a long run of commercial operation. Moreover, the cost of pollution is not yet charged against the power producers. It will not be easy to convert industry to supply and distribute alternative fuels such as methanol or hydrogen for transport. But by so doing, the pollutant emissions from road vehicles and aircraft could be greatly reduced, even eliminated.

Understanding the living Earth

It was the view from space about 20 years ago that showed us how beautiful and how seemly is our planet when seen in its entirety. Above all, this vision reminded us that 70 per cent of the Earth's surface is ocean, something we often forget because of our obsession with human affairs on the land surface. As this book has shown, we are only just beginning to glimpse the extent to which the world's oceans and the marine life in them are important in regulating the climate and the chemistry of the Earth. Marine biology is a cinderella among sciences, particularly in this country. Yet it was the pioneering work of my colleagues Patrick Holligan, Andrew Watson, Mike Whitfield, and Toby Tyrrell at the Marine Biological Association of the United Kingdom that first demonstrated the planetary significance of microscopic algae living at the ocean surface (see Chapters 6 and 7).

The algal life of the oceans is important for climate in three possible ways. Firstly, they emit sulphur gases that oxidize to form cloud condensation nuclei. The clouds over the oceans, with their net cooling effect, may be a consequence. The second discovery concerning marine algae came from the observations made by British scientists in the North Atlantic in 1991. They found that the great springtime blooms of the diatoms covering areas of millions of square kilometres use up the carbon dioxide in the sea surface so that its concentration falls to a level far below that in equilibrium with the atmosphere. They serve as a powerful pump that removes carbon dioxide from the air. The third way in which algae can affect climate is simply by their presence or absence. In their absence, the ocean water is so clear that sunlight penetrates deep and does not heat the surface layers. Conversely, dense algal growth absorbs and scatters sunlight and causes surface warming in the areas where it is found.

Another region of the Earth's surface that could have a significant climatic role is the rainforest of the humid tropics. As we saw in Chapter 8, humankind is fast destroying these forests. Yet in spite of optimistic signs coming from Brazil, and Columbia too, we are still removing tropical forests at a ruthless pace. And often when the forests go, the rain goes with them. Their replacement by crude cattle farming could precipitate a disaster for the billions of the poor in tropical regions. It is heartbreaking to conceive of the human suffering, the guilt, and the political consequences that would be triggered by a refugee problem involving a major part of the humid tropics. The change in climate when the forest went would also have secondary climatic consequences for the temperate regions.

That the danger of climate change from forest clearance is real was illustrated in an unusual television documentary about the Panama Canal some years ago. The history of this amazing feat of engineering was used to illustrate a new threat to its continued function. The threat was not, as you might imagine, from local politics, but from agriculture. The canal climbs over the isthmus of Panama through a series of locks. The entire system is powered and filled by the abundant rainfall of that humid region. But the rain and the trees of the forests are part of a single system. Now that the forests are being destroyed to make cattle ranches the rain is declining and may become too little to sustain and power the canal. I hope somehow the fact that this great work of human engineering is threatened by our insatiable desire for beef, will serve to bring home the consequences of our actions in burning away the forests of the Earth.

As we have seen, a planetary physician would regard the great forests of the tropics as part of the skin of the Earth; like human skin they sweat to keep us cool. Sweating is part of our personal refrigeration system. The evaporation of water from the forests is part of Gaia's cooling system. It works because the white clouds that persist over the humid tropics have a net cooling effect. They cool by reflecting back to space sunlight that would otherwise reach the ground and add its increment of heat to what appears to be an overheated planet. Clouds also reflect back to the ground heat that might otherwise escape to space, but their net effect is usually to cool.

Placing a value on the forests

Brazilian scientists were once asked by their government to calculate the value of the forests of the Amazon as producers of oxygen for the world. The government spokesman argued that without the oxygen their trees produced, fuels like coal and oil would be worthless. Some charge should therefore be made for the export of the essential gas, oxygen. It was a fine idea, but unfortunately calculations of the net production of oxygen by the forest gave an answer close to zero. The animals and microorganisms used up almost all of the oxygen the trees produced.

Amazonia may not be worth much as a source of oxygen, or by the same calculation, as a sink for carbon dioxide, but it is a magnificent air conditioner, not only for itself but also for the world through its ability to offset, to some extent, the consequences of greenhouse gas warming. Do the forests then have an estimable economic value as natural regional, if not global, air conditioners?

One way to value the forests as air conditioners would be to assess the annual energy cost of achieving the same amount of cooling mechanically. If the clouds made by the forests are taken to reduce the heat flux of sunlight received within their canopies by only 1 per cent, then their cooling effect would require a refrigerator with a cooling power of 6 kilowatts per hectare. The energy needed, assuming complete efficiency and no capital outlay, would cost annually, $1300 per hectare. How does the value of this freely given benefit compare with that of land cleared for cattle raising? That is the usual fate of land in the humid tropics.

A hectare of cleared tropical forest is said to yield enough meat for about 1850 beefburgers annually. At the site the meat is not worth more than about $40, using monetary values of the 1990s, and this only during the very few years that the land can support livestock. Unlike on cleared land in the temperate regions, beef production cannot be sustained in the tropics and the land soon degenerates to scrub or even desert. Next time you eat a burger, or watch someone else eating one, think of the real cost of its production, the stripping of an asset worth $65. Yes, the five square metres of land needed to produce enough meat for one burger has lost the world a refrigeration service worth about $65. On this basis a reasonable estimate of the worth of the refrigeration system that is the whole of Amazonia is about $150 trillion.

Such a valuation in terms of the refrigeration capacity of the trees alone is an underestimate. Just now the forests sustain a home, a habitat for vast numbers of organisms including a billion people around the Earth. The forests are more valuable to us all than we have yet grasped; like love itself, they are so valuable that we take them for granted.

The legend of Atlantis

Common sense now tells us that in the absence of a clear understanding of the consequences of what we are doing to the Earth we should cut back our pollutions and land abuse to the point where at least there was no annual increase. But like all acts of self-denial it is only too easy to put it off until something happens. I can't arrange, or predict, anything exciting enough to cause us to give up polluting, but what I can do is to tell you a fable about an environmental problem that afflicted an imaginary industrial civilization 15,000 years ago.

Just about the time that our immediate ancestors appeared on the Earth, 2.5 million years ago, the planet itself was changing from a state where the climate was comparatively constant to one where the climate cycled periodically between glacial and interglacial phases. The ice ages were long; they lasted some 90,000 years. In the intensely cold winters, ice extended to within 45 degrees latitude of the equator. The warm periods, the interglacials, were brief, lasting only about 12,000 years, and the climate was like the one we now enjoy (or at least did until quite recently).

I would like you to imagine that civilization became industrial 15,000 years earlier than it did. This requires an increase of only 0.5 per cent in the rate of evolution of human society. Imagine a civilization very like ours now, just as developed and just as polluted, with meetings to discuss the consequences. The main difference is that the Earth was then in an ice age. The climate of London was like the present climate of Iceland and there were few inhabitants. The oceans were more than 130 metres lower than now. A vast area of land, mostly near the equator and now under the ocean, was dry and populated.

Let's imagine that the civilization developed and became industrial somewhere in the region of Japan and China. The cold winters with the need for housing and heating stimulated invention. The region was also rich in coal, oil, and mineral deposits. Soon these were exploited and there followed a rapid progression through water and wind power to steam, electricity, and nuclear power, just as we ourselves have seen take place.

Greenhouse gases, carbon dioxide and methane from extensive agriculture, began to rise and before long the climate became perceptibly warmer. A large part of the civilized world was in the tropical zones and these were becoming uncomfortably warm for their inhabitants. The nations of the north were efficient at producing consumer goods. They were like the Victorian British, or the present-day Japanese. Needs drive invention, and soon refrigerators using CFCs were pouring from the production lines of the island of Atlantis.

It was not long before scientists began to realize that the global environment was changing. A few of them stumbled on the fact that CFCs leaking from refrigerators were accumulating in the air without any apparent means for their removal. Soon it was discovered that CFCs were a threat to the ozone layer and that if their growth in abundance in the air continued, ozone in the stratosphere would be so depleted that many, especially the fair-skinned, would be in grave danger from the ultraviolet component of sunlight and would develop skin cancers. There was an explosion of hype in the media over this threat and funds flowed for science as never before. Governments were reluctant to act because they knew the CFCs to be harmless in the home, efficient refrigerant gases, and the basis of a large and profitable industry. They were reluctant also because there was no evidence of any increase in solar ultraviolet at ground level; indeed there was a decrease. So nothing was done to stop CFCs rising in abundance at 10 per cent each year.

A few scientists felt frustrated because they knew that the real threat from CFCs was not ozone depletion but their properties as greenhouse gases. The fear of cancer always seems to transcend other dangers and as a consequence ozone depletion was the issue that received the most attention. Greenhouse warmth was known about but regarded as a good thing, since the world was cold anyway.

Nobody then knew that in two thousand years the planet was destined to make one of its characteristic jumps in temperature to an interglacial. The polar regions in the ice age were so inhospitable that there was none of the clear strong data from ice cores that we now have to help us understand the past. The jump in temperature due 2000 years ahead would come because the position of the Earth with respect to the Sun was changing in a way that increased the heat received from the Sun. These were small increases in solar heating and by themselves insufficient to precipitate an interglacial. But at the end of a glacial period the planet was in a state highly sensitive to small perturbations.

A minority opinion among the scientists held that the increasing warmth from greenhouse gas pollution would start to melt the ice caps and that the flooding of low-lying tropical forest land would then take place. This in turn would release vast volumes of methane gas as the vegetation rotted beneath a few metres of sea water. The methane would cause more greenhouse warming and soon by a runaway positive feedback the planet would heat and melt the vast polar ice caps. They warned that the rise in sea level would ultimately be 150 metres, enough to drown most of the large towns and cities of the civilized world.

This pronouncement was treated with derision and contumely. Ozone depletion, and the dangers from nuclear power, were the main interest to government and environmentalists alike. Soon the CFCs reached 5 parts per billion in the air and ozone holes appeared over the poles. By themselves the ozone holes were of no consequence since nothing lived at the ice-age polar regions. But their presence was enough to tip the balance in favour of legislation to ban the use of CFCs. Unfortunately it was too late, for the greenhouse balance had also tipped and the planet was now like a boat passing over the edge of a waterfall, moving ever faster towards the heat of the interglacial. The polar ice was already melting and within a few hundred years all of this Atlantean civilization was deep under the ocean. The legend of a flood and of a great empire beneath the ocean persisted. The stories about it were reiterated over the camp fires of the wandering tribes of hunters from generation to generation.

If there is any moral to be drawn from this tale it is that we are very lucky to have chosen to pollute the air now, when the planet is least sensitive to perturbation by greenhouse gases. But if you look at the Earth as I do, as a superorganism, then we need to make sure that some other surprise may not be waiting to do to us unexpected damage. A surprise as great at that which confronted those imaginary Atlanteans. So let me conclude with some further thoughts about the dangerous illusion that we humans could be stewards of the spaceship Earth.

Could we, by some act of common will, change our natures and become proper stewards, gentle gardeners taking care of all the natural life of our planet? I think that we are full of hubris even to ask such a question, or to think of our job description as stewards of the Earth. We are all too plainly failing even to manage ourselves and our own institutions. I would sooner expect a goat to succeed as a gardener than expect humans to become responsible stewards of the Earth.

Originally a steward was a "styward", the keeper of the sty where the pigs lived. This was too lowly a post for most humans, and gentility raised it to that of steward, who thus became a bureaucrat, in charge of men as well as pigs. Do we really want to be the bureaucrats of the Earth? Do we want the full responsibility for its care and health? There can be no worse fate for people than to be conscripted for such a hopeless task – to be made forever accountable for the smooth running of the climate, the composition of the oceans, the air, and the soil. Something that, until we began to dismantle creation, was the free gift of Gaia.

I would suggest that our real role as stewards of the Earth is more like that of that proud trades union functionary, the shop steward. We are not managers or masters of the Earth, we are just shop stewards, workers chosen, because of our intelligence, as representatives for the others, the rest of life of our planet. Our union represents the bacteria, the fungi, and the slime moulds as well as the nouveau riche fish, birds, and animals and the landed establishment of noble trees and their lesser plants. Indeed all living things are members of our union and they are angry at the diabolical liberties taken with their planet and their lives by people. People should be living in union with the other members, not exploiting them and their habitats. A planetary physician observing the misery we inflict upon them and upon ourselves would support the shop steward and warn that we must learn to live with the Earth in partnership. Otherwise the rest of creation will, as part of Gaia, unconsciously move the Earth itself to a new state, one where we human may no longer be welcome.

Glossary

aeronomist an atmospheric scientist. Word coined by Sydney Chapman, who discovered the ozone layer

aerosol a dispersion of fine liquid droplets or solid particles, usually in air, but can be in any gas

albedo the proportions of light reflected from a planet, going from an albedo of 1 for complete reflection to 0 for no reflection

anaerobe microorganism that can live only in the complete absence of oxygen

anoxic having no oxygen

antioxidant substance that protects against destruction by oxygen

Archean geological eon, when life started (see table)

argon wholly unreactive gas making up nearly 1% of the air

asthenosphere layer of plastic rock 70–250km below the Earth's surface

astrophysicist a scientist who is mainly interested in stars and other heavenly bodies, how they work, and their structure

atmosphere the least massive, yet the most important part of the Earth for life. Through the atmosphere pass nearly all of the elements that go to form living organisms. The atmosphere protects life from the rigours of space and establishes the climate

atmospheric pressure the pressure the atmosphere exerts on the Earth

bacteria the first organisms to colonize the Earth as a whole, they were the whole of life on Earth for over a billion years and still are omnipresent. They exist as small spherical or cylindrical cells about one micron in diameter that have no nucleus and multiply by division

basalt a lava rock representative of the magma, and the original material of the Earth. It tends to be alkaline and reduced, in contrast with the other principal igneous rock, granite, which is acidic and oxidized

becquerel a recent unit of radioactivity named after the discoverer of radioactivity. One becquerel is one disintegration per second. It replaces the curie, which is inconveniently large at 37 billion disintegrations per second

betaine an electrically neutral salt that does not harm cells at high concentrations as would ordinary salt

biodegradable can be broken down into harmless products by bacteria or other biological processes

biogeochemistry the branch of science concerned with the chemistry of living organisms on a planetary scale. Particularly the cycling of the important elements of life – carbon, nitrogen, and sulphur – through the atmosphere, rocks and oceans

biosphere the realm of all living organisms

biota all the life on Earth. The totality of organisms from viruses to giant redwood trees, from bacteria to whales

carbon dioxide, CO_2 a gas with a faintly pungent smell, present in the air at only 280 parts per million, but in your breath at 4%. Used by plants, produced by respiration and burning. Carbon dioxide in the air helps, through the greenhouse effect, to keep the Earth warm, but too much may lead to overheating

carcinogen substance or radiation that if administered increases the probability of the eventual development of cancer

centrioles root-like structures found in pairs in animal cells (not plant cells) that form the ends of the spindle during cell division

chemical equilibrium the state reached when all of the chemical energy of a system has been released

chemically inert describes a substance that resists all or most attempts to change it chemically, e.g. argon gas

chemical potential energy the energy stored in chemical bonds or available from reactions, see p. 117

chloroplast the organelle containing chlorophyll in plants cells

coccus spherical bacterium. Staphyllococci and streptococci include pathogenic species

coccolithophores marine algae that make their shells from calcium carbonate

Coriolis force the force experienced by any material thing when it moves along a line of longitude. Because the Earth rotates, any object moving between a pole and the equator will

experience an acceleration due to the Coriolis force

creationist person who believes the Earth and all life is the work of a Creator, and dismisses evolution theory

Cretaceous geological period (see table) distinguished by the sedimentation of white, chalky algal tests

Cretaceous/Tertiary (K/T) boundary a thin rock zone, free of fossils, deposited between the end of the Cretaceous and the beginning of the Tertiary periods

cyanobacteria simple blue-green photosynthetic bacteria with an ancestry reaching as far back as the first life on Earth

cybernetics the science that expresses control theory. Important in self-regulating systems of electronics and mechanical design and in the physiological systems of living organisms and of superorganisms such as Gaia

diatoms unicellular algae whose cells are formed from two overlapping halves. Diatoms possess exquisitely beautiful silica skeletons

diffusion the mixing of different substances through the random motion of their atoms, molecules and ions

DNA deoxyribonucleic acid. The natural polymeric material that carries all of the information needed and used by living organisms

electrically neutral having no net positive or negative charge

emergent property an entirely new property that appears as a result of the evolution of a system. Self-regulation is an emergent property of some operating devices and all living organisms

endosymbiosis hypothesis see p. 102

eukaryote an organism whose genetic material is carried in the nucleus of its cell. All cells other than bacteria are eukaryotic

euphotic zone the top layer of a body of water in which there is light enough for photosynthesis

eutrophic describes a sea or lake with surplus nutrients and high rate of photosynthesis, which stimulates excessive algal growth. When the algae die they decompose, depleting the oxygen so the aquatic animals die

fermenters bacteria that live on chemical energy from the excreta or dead bodies of primary producers

fixation the biochemical act by which a plant takes in atmospheric CO_2, water, and nitrogen, and forms organic matter using the free energy of sunlight

free radical a molecule or an atom that has one or more electrons unpaired. A highly reactive molecular fragment formed during a chemical reaction

Gaia theory sees the Earth as a system where the evolution of the organisms is tightly coupled to the evolution of their environment. Self-regulation of climate and chemical composition are emergent properties of the system. The theory has a mathematical basis in the model "Daisyworld"

geochemist chemist interested in the composition and reactions of the materials of the Earth, including the atmosphere, oceans, and rocks

geophysicist physicist interested in the dynamic physical state of the Earth and its compartments: the atmosphere, oceans, and rocks. Geophysics embraces meteorology and climatology, and their oceanic equivalents, plus the study of the form and dynamics of the solid and molten parts of the Earth

geosphere a dubious term supposedly referring to all of the Earth that is non-living

gene the part of a chromosome that carries the coded information for hereditary characteristics

Hadean the Earth's first and most violent geological eon (see table)

Hadley Cell large circulating mass of tropospheric air formed by uprising warm air and its down flow in regions near the poles. Named after the meteorologist, Hadley

halophile bacterium bacterium that can live and reproduce in strong salt solutions

heterotroph organism that gains energy from the oxidation of organic matter. Includes all animals

homeostasis that wisdom of the body whereby a state of constancy is kept in spite of external or internal environmental change. Our temperature, blood salinity, and acidity are kept constant by homeostasis

hydroxyl radical compound of hydrogen and oxygen with an unpaired free electron; one of the most powerful oxydizing reagents known. Present in

air, and the means for the removal of almost all organic gases, such as methane, from the air

ideal gas a model for a hypothetical gas that behaves "perfectly" i.e. obeys all the gas laws, which relate the temperature, volume, and pressure of a gas

infrared radiation electromagnetic radiation at the wavelengths emitted by a red-hot body, longer than the wavelengths of visible light

inorganic the opposite of organic. Inorganic chemistry is the chemistry of non-living matter

ion an atom with a positive or negative charge due to having lost or gained electrons respectively

ionization the process of forming ions. Atoms can lose or gain electrons in collisions, in solution, or as a result of radiation

ionosphere the upper layer of the air where some of the gas molecules have been ionized by high energy radiation or atomic particles from the Sun. Older forms of radio communication depended upon its presence for transmission well beyond the horizon

isotopes atoms of an element which differ in weight but have the same atomic number. Isotopes of an element are almost indistinguishable chemically

Jurassic geological period (see table)

lipid organic compound found in living organisms. Usually oily or waxy and made mostly of carbon and hydrogen but with a little oxygen and sometimes nitrogen also. Lipids are usually, but not in every case, water repellent

lithosphere the Earth's crust

logarithmic scale a graph scale where each unit represents a ten-fold increase in the quantity

luminosity the measure of brightness of a star: the amount of energy radiated per second

magma hot plastic rock beneath the Earth's crust. It comes to the Earth's surface as volcanic lava

marasmus somewhat obsolete term for a wasting disease

megaton a million tons

mesosphere layer of atmosphere between the stratosphere and thermosphere

meteorologist scientist who studies the Earth's atmosphere, especially with regard to weather

methane, CH_4 natural gas. Comes mainly from the fermentation of

organic detritus in the absence of oxygen. An essential part of the natural cycle of carbon and for the continuation of life

methanogens bacteria that convert organic matter into methane and carbon dioxide

Milankovich effect see p. 149

mitochondrion an organelle within a eukaryotic cell, with its own independent genes, that conducts the oxidative energy transactions of the cell

multicellular organism organism made up of more than one cell

mutagen substance that causes genetic mutation

neo-Darwinism theory of evolution that combines genetics and Darwin's theory of natural selection

neoplasm abnormal growth of cells forming a tumour

noble gases the monatomic gases helium, neon, argon, krypton, xenon, and radon

nucleate to form a nucleus

organelles the mitochondria, nuclei, and lysosomes in plant and animal cells

organic composed of carbon compounds. Mainly matter that is or has been alive, but also some synthetics made from carbon compounds

organism living plant or animal, including microbes

osmotic pressure pressure required to stop the flow of solvent from a solution by osmosis, i.e. through a semipermeable membrane separating two different solutions

outgas to liberate gas from a planet's surface to its atmosphere

oxidation chemical reaction where a substance loses electrons. Often the substance also joins with oxygen

oxidizing bringing about oxidation

ozone layer a layer of air 15–50 km above the Earth's surface, where most atmospheric ozone is found. It absorbs some of the ultraviolet radiation of the Sun, and in so doing grows warmer and serves to lessen the ultraviolet received at the surface. It is now being depleted to some extent by a complex sequence of reactions involving manmade emissions from the Earth's surface of methane, chlorine, and nitrogen compounds

paleobotanist scientist who studies fossil remains of plants

Pangaea large continent that broke up 200 million years ago. See p. 52

peptide bond see p. 117

Permian geological period (see table)

Phanerozoic geological eon (see table)

photosynthesis the process in which plants use the energy from sunlight to synthesize organic compounds for their growth from carbon dioxide in the air and water. The process also produces oxygen

planetesimal a small planet, or piece of rock, 1–50km in diameter. These orbit the sun and occasionally collide with the Earth and other planets

plate tectonics the theory that the Earth's surface is made up of plates, which have moved during geological history to form the present land masses and oceans

prokaryote organisms, e.g. bacterium, whose genetic material is not contained in a cell nucleus

Proterozoic geological eon (see table)

protist microbe composed of a single nucleated cell, e.g. alga, fungus

quark an elementary particle: the fundamental constituent of all matter

radioactivity the spontaneous release of energy in the form of high energy particles or radiation from atomic nuclei. Some elements essential for life, such as potassium, carbon, and hydrogen, have naturally occurring isotopes that are radioactive

radio-frequency radiation electro-magnetic radiation at low frequencies and long wavelengths compared with light or heat radiation. Generated naturally by lightning and stars, but on the Earth overwhelmingly more by humans for communication and cooking

redox potential a chemist's measure of the tendency of a system for oxidation or reduction

reduction a chemical reaction where an atom or molecule gains electrons, often gaining a hydrogen atom or losing an oxygen atom

reducing substance substance that can combine with oxygen, such as a food or a fuel. More strictly a substance that donates electrons in a chemical reaction

reductionist theory theory that attempts to explain a complex system by analyzing its simplest components

reflectivity the ability of a surface to reflect radiation incident upon it

regolith the mixture of dust, pebbles and rocks at the surface of a dead planet. In great contrast to the soil of the Earth, which is a rich ecosystem of organisms and their detritus

resonate to oscillate at the natural frequency of a system, often producing large amplitude vibrations. To be in tune with

sarcina types of spherical bacteria that aggregate in cubical packages like the atoms of a crystal

solute the substance dissolved in a solvent in a solution

spirochetes thin motile corkscrew-shaped bacteria

staphylococci types of spherical bacteria bunched together like grapes

stratosphere layer of the atmosphere immediately above the troposphere where the air is stratified. It is warmer than the upper troposphere and contains the ozone layer

streptococci types of spherical bacteria connected like a chain of beads

stromatolyte stony mass formed from calcium carbonate secreted by cyanobacteria

supernova a star that explodes with such violence that for a period of months as much light is emitted as from all the stars of a galaxy. In one kind of supernova heavy elements, e.g. uranium and iron, are synthesized by nuclear processes and distributed into space

sulphur cycle see p. 125

symbiont an organism living in symbiosis with another

symbiosis when two or more species live together, to the benefit of each other

tectonic relating to movements in the Earth's crust

thermosphere the outermost layer of the Earth's atmosphere beyond 90km

trace gases gases such as methane, nitrous oxide, ozone, carbon monoxide, dimethyl sulphide, and methyl chloride that exist at low concentrations in the atmosphere. Their low concentration belies their importance

troposphere the layer of the Earth's atmosphere from the surface to about 10km, where the air is well mixed, where birds fly, and where clouds are mostly to be found

ultraviolet radiation electro-magnetic radiation in the range of frequencies immediately above visible light and below X-rays. The ultraviolet reaching the Earth's surface has both harmful and beneficial effects

urea the product of nitrogen excretion in animals

Van der Waal's force the weak attractive force between atoms or molecules

vitalism the belief that living matter has a "soul" or "spirit" which distinguishes it from non-living matter

Geological timescale

Millions of years ago	Eon	Period
4600	Hadean	Pre-Cambrian
3700	Archean	
2500	Proterozoic	
700	Phanerozoic	
570		Cambrian
500		Ordovician
440		Silurian
395		Devonian
345		Carboniferous
280		Permian
225		Triassic
190		Jurassic
140		Cretaceous
65		Tertiary
1.8		Quaternary

190

Index

bold figures indicate main entries; *italic* indicate illustrations